ACOUSTICS AND THE BUILT ENVIRONMENT

ACOUSTICS AND THE BUILT ENVIRONMENT

ANITA LAWRENCE

*Graduate School of the Built Environment,
The University of New South Wales,
Australia*

ELSEVIER APPLIED SCIENCE
LONDON and NEW YORK

ELSEVIER SCIENCE PUBLISHERS LTD
Crown House, Linton Road, Barking, Essex IG11 8JU, England

Sole Distributor in the USA and Canada
ELSEVIER SCIENCE PUBLISHING CO., INC.
655 Avenue of the Americas,
New York, NY 10010, USA

WITH 23 TABLES AND 59 ILLUSTRATIONS

© 1989 ELSEVIER SCIENCE PUBLISHERS LTD

British Library Cataloguing in Publication Data
Lawrence, Anita
    Acoustics and the built environment.
    1. Buildings. Acoustics, — For architectural design
    I. Title
    729′.29

    ISBN 1-85166-308-8

Library of Congress Cataloging-in-Publication Data
Lawrence, Anita.
    Acoustics and the built environment/Anita Lawrence.
    p.   cm.
    Includes bibliographies and index.
    ISBN 1-85166-308-8 (U.S.)
    1. Soundproofing. 2. Buildings—Vibration. 3. Acoustical
engineering. I. Title
TH1725.L35 1989
693.8′34—dc 19

Printed in Great Britain by Page Bros (Norwich) Ltd

Preface

People have five senses to guide them through their interactions with their environment, although usually one or more of the senses will assume the greatest importance at any given time. In a summer garden, for example, the visual, olfactory and thermal senses may tend to dominate a person's reactions; however, the tactile sense may be stimulated by a rose thorn and the aural sense may be overwhelmed by a low-flying aircraft.

This book is primarily concerned with the aural environment, which is very important to people's well-being. However, it cannot be considered in isolation: for example, when wanted sounds are heard, when listening to music or poetry, or even when taking part in ordinary conversation, aesthetic and emotional factors will also affect one's acoustic response. When talking or listening to speech, either directly or through electronic means, the most important consideration naturally is intelligibility—and this is strongly affected by the acoustic environment in which the communication is taking place. When relaxing, or trying to sleep or to concentrate on some activity, unwanted sound can cause annoyance, which may in some cases lead to stress. In the worst case, noise levels may be sufficiently high to cause permanent, irreversible deterioration of hearing acuity.

It is therefore very important that the professionals involved with the design and construction of the 'built environment' should understand the principles of acoustics. For example, unlike impure air and water, which may affect communities far from the source of the pollution, the effects of noise are relatively confined; thus regional planners can prevent noise annoyance by ensuring that incompatible land uses are not located close to major noise sources such as transportation routes, airports, etc. At both regional and local planning scales, noise is an important consideration when assessing the environmental impact of any new development. Topographical and meteorological influences on the propagation of sound out of doors must also be understood.

At the level of the individual building or building complex, the archi-

v

tect needs to consider external and internal sound sources and the relative noise tolerance of different building uses at the initial site planning stage as well as when carrying out detailed planning of the building(s). Since in the real world, incompatible land uses do occur, it may be necessary to use special building envelope elements to ensure that at least inside the building, the acoustic environment will be acceptable. Conversely, if the building is a potential noise source, e.g. an industrial building located close to a residential area, both planning and selection of building envelope elements will be required to prevent unfavourable community response. Transmission of noise from one part of the building to another must be understood and controlled. In the case of certain building types, such as auditoria, concert halls, lecture rooms, etc. detailed analyses of the behaviour of sound within enclosed spaces are required, which will necessitate the careful selection of room shape and room surfacing elements.

Mechanical services engineers also have an important role to play in the control of noise and vibration within buildings, and unless they also understand and apply acoustic principles, excessively high noise levels from air-conditioning systems, plumbing and the like will result. Such faults are extremely difficult to rectify in a completed building.

Once the broad choices regarding constructional elements have been made, the acoustic performance of a building is critically dependent on construction detailing and on the faithful execution of these details by the contractor and the building tradesmen. Informed supervision is thus essential during the building process, as is the education of the construction team regarding the importance of exact compliance with the detailing. For example, a massive masonry dividing wall may have been selected to provide good sound isolation between two rooms; the additional costs of such construction will be wasted if flanking transmission occurs through small air gaps, or through the ceiling space above. Sometimes it is necessary to use a 'room-within-a-room' construction to obtain isolation from structure borne sound in rooms such as studios; this type of construction needs very careful detailing and constant supervision, because the whole system depends on there being no structural connection between the spaces; even a fallen brick bridging the cavity can allow audible sound to be transmitted. Often such mistakes are difficult to identify or to remedy, once the building has been completed.

Increasingly, acoustical criteria are included in planning and building legislation, and in some cases, acoustical measurements using sophisticated equipment are required by the authorities to demonstrate compliance. In addition, although considerable research has provided useful design guides for predicting sound propagation out of doors, taking into account topography and meteorological conditions and the data base regarding the acoustical characteristics of building materials and elements is growing, there is often the need for acoustical measurements to be made, either in the laboratory or *in situ*. The International Standards Organisation (ISO) and the International Electrotechnical Commission (IEC) have now published many acoustic standards. These cover the measurement of such factors as the performance of building materials; the acoustical acceptability of buildings, including noise transmission between spaces and the acoustical characteristics of auditoriums, offices, etc.; the noise emission of aircraft, trains and motor vehicles and road traffic; the noise emission of machines and equipment; community noise measurement and assessment; and also the requirements for acoustical instrumentation. Many countries have also published their own national acoustical standards, based for the most part, on the ISO and IEC documents. These standards are frequently referred to in legislation, in which case, of course, compliance is mandatory.

As mentioned above, acoustical problems are usually very expensive, and frequently impossible to correct. It is essential that the various professionals responsible for planning, designing, building and assessing the built environment have sufficient knowledge and understanding of acoustics to avoid such problems occurring. In these complicated times it is not expected that every architect, planner, engineer or contractor will be an expert in acoustics; however, it is essential that each understands enough of the subject at least to know when specialist advice should be sought. Although it is a relatively new profession, there are now many acoustic experts available with the knowledge required to assist the other professionals concerned with the built environment, and the latter will be negligent of their client's interests if they do not seek acoustic advice. It is the intention of this book to assist the 'built environment' professionals themselves to understand the important principles of acoustics that affect their work.

The material has been arranged to enable direct reference to particular items of interest. Chapter 1 deals with the perception of sound and vibration by people and also with the basic physics of sound and sound sources. Chapter 2, Noise in the Community, is specifically directed to planners and environmental consultants; it covers community noise sources, including transportation, guidelines for assessing acceptable noise levels and noise propagation models. Chapter 3 deals with specific planning concerns, such as guidelines for the siting of major transportation routes and terminals, industrial areas, etc. This chapter is cross-referenced to the earlier chapters for more detailed explanations; however it may be the first source of information for the planner with a specific problem to solve.

Chapters 4 and 5 are primarily directed to architects, builders and services engineers. Chapter 4, on room acoustics, is concerned with the characteristics of wanted sound, such as speech and music, propagation of sound in enclosed spaces, and sound absorbent materials. Chapter 5 deals with the transmission and control of sound in buildings. This is followed by Chapter 6 which deals with specific building types. This chapter may be consulted directly by the designer of a particular building as it outlines the main factors to be taken into account in each case; it is cross-referenced to earlier chapters for specific details.

Three Appendices include some information on the measurement of sound absorption coefficients, and data is also included regarding the acoustical performance of common building materials and systems.

A comprehensive list of References is included for each chapter, to enable readers to obtain more detailed information on specific topics than could be included in this relatively slim volume.

Anita Lawrence,
Graduate School of the Built Environment,
University of New South Wales,
P.O.Box 1, Kensington, N.S.W. 2033,
Australia.

Contents

CHAPTER 1

Sound, Vibration and Human Perception

1.1 INTRODUCTION

All sound has as its origin the mechanical vibration of the particles of a medium; however, it is common to differentiate between vibration that is *heard* as sound by reception through the ears, and sound that is *felt* through various other receptors in the body.

Sound does not form part of the electromagnetic spectrum (which includes radio, light and heat waves). Electromagnetic waves travel so quickly that energy transference is perceived by humans to be instantaneous. Sound energy travels relatively slowly, which results in some important practical effects. For example, in a concert hall, if the shape is incorrectly designed, echoes may be clearly heard by the audience or by the performers, due to the perceptible difference in the arrival times of sound that has travelled directly from the orchestra to the listener and the sound that has travelled via a series of reflected paths from the walls, ceiling and other surfaces and objects.

This relatively slow propagation of sound energy in air means that the physical wavelengths corresponding to frequencies in the audio range extend from about 17 mm (3/4 in) to 17 m (55 ft). When a surface or an object is similar in size to, or smaller than the wavelength of the impinging energy, diffraction takes place and the energy is scattered, rather than being reflected specularly, as light is reflected in a mirror. This means in practice that it is difficult to determine the propagation paths of low frequency (long wavelength) sounds within buildings; it also affects the accuracy of prediction of the effectiveness of roadside barriers and topography in reducing traffic noise, for example.

1

In this chapter a brief description will be found of the physical characteristics of simple sound sources and sound propagation in air; acoustic descriptors are defined and the response of people to sound and vibration is discussed.

1.2 PHYSICAL CHARACTERISTICS OF SOUND

A sound wave is produced when a medium is set into vibration by a source. The medium may be gaseous, liquid or solid. The simplest sound source is called a 'point' source, which may be represented as a small pulsating sphere located in a gaseous medium such as air. As the sphere alternately increases and decreases in diameter, the surrounding air will be subjected to alternating local increases and decreases in pressure, compared to the undisturbed condition, or atmospheric pressure. When the sphere diameter increases, a small *compression* of the gas particles occurs in the immediate vicinity of the source. This increased pressure means that there is now a pressure differential between these particles and those of the surrounding air. This excess pressure is transferred spherically outwards from the source. Conversely, when the sphere contracts, a slight vacuum, or *rarefaction*, will be formed in the immediate vicinity of the source, and this will also be transferred outwards. As the pulsation continues, there will be a series of compressions and rarefactions propagated on a spherical wavefront (see Fig. 1.1).

The distance between two successive compressions depends on two factors: a) the rate at which the pulsation take place and b) the rate of transfer of energy through the medium. The first factor is called the *frequency* of the sound, and it is measured in terms of the number of complete cycles of oscillation per second; the unit is the *Hertz*, abbreviated as Hz, and the symbol used is f. The second factor is the *velocity* of sound in the medium, measured in metres per second (or feet per second) and the symbol used is c. The distance between two successive compressions (or rarefactions, or any other particles undergoing the same *phase* of vibration) is called the *wavelength* of the sound; it is measured in m (or feet) and the symbol used is the Greek letter λ.

The relationship between these three units is:

$$c = f\lambda \qquad\qquad\qquad [1.1]$$

The speed of sound, i.e. the rate of transfer of energy through the medium, occurs more quickly through a medium that is dense than

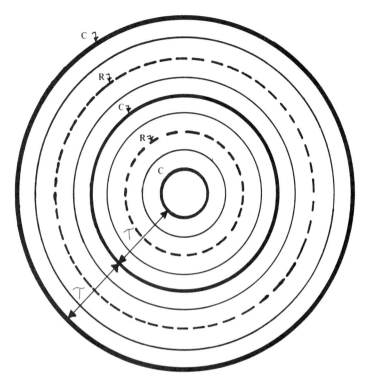

Fig. 1.1. Sound wave propagation from a point source in a free field; thick solid lines = compression, C; dotted lines = rarefaction, R; thin solid lines = undisturbed, atmospheric pressure; wavelength = λ.

through one that is not. Thus the speed of sound is greatest in solid media, such as concrete, lower through liquids and slowest through gases such as air. In air, at normal room temperatures and pressures, the sound velocity is approximately 344 m/s (1130 ft/s).

The motion undergone by an individual particle affected by a sound wave emitted by a point source in air will now be considered. The particle will first be displaced in the direction of propagation of the sound energy, away from the source, due to the compression caused by the outward expansion of the sphere, to a maximum positive displacement, of amplitude ξ. It will then be drawn back towards the source (as far as the maximum negative displacement) under the influence of the rarefaction caused by the decreased diameter of the sphere. As

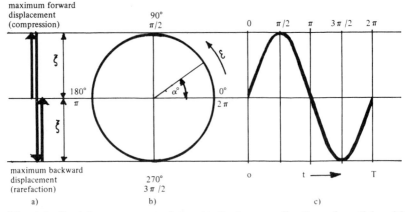

Fig. 1.2. Particle movement: a) longitudinal wave vibration of particle with maximum displacement ξ (shifted sideways for clarity); b) projection of longitudinal vibration on to reference circle, angular velocity $\omega = 2\pi f$ radians/second; c) representation of one cycle of particle vibration as a time displacement curve; phase angle $\alpha°$ in radians also shown.

mentioned earlier, the repetition rate of this forward and backward oscillation about the particle's rest position is called the frequency, f, of the source; thus if f = 100 Hz, the particle will undergo 100 complete oscillation cycles per second. In order to visualize the particle movement more readily it is usual to project it onto a reference circle (see Fig 1.2.a) and 1.2.b)). The particle commences at its undisturbed location, designated 0°, and then moves to 90° (or $\pi/2$) at maximum positive displacement, back to 180° (or π) when it passes through its rest position; then to 270° or ($3\pi/2$) at maximum negative displacement and thence to 360° (or 2π) at the completion of one cycle. The complete oscillation cycle corresponds to an angular rotation of 2π radians, and the angular velocity of the particle, ω, is equal to 2π divided by the time taken to complete the cycle, T: thus

$$\omega = 2\pi/T \qquad\qquad\qquad [1.2]$$

or, since f = 1/T,

$$\omega = 2\pi f \qquad\qquad\qquad [1.3]$$

Alternatively, the particle displacement may be shown against time, as in Fig. 1.2.c) which shows the familiar sine wave of Simple Harmonic Motion, the equation for which is

$$y_i = \xi \sin \omega t \qquad [1.4]$$

where

y_i = instantaneous particle displacement,
ξ = amplitude of maximum particle displacement
t = time.

Particle velocity, v_i, is the rate of change of displacement with time, (dy/dt), thus

$$v_i = \xi \cos \omega t \qquad [1.5]$$

As the sound wave passes through the air, the *pressure*, measured in Pascals, symbol p, increases and decreases with respect to atmospheric pressure. In a freely moving sound wave, the ratio of sound pressure to instantaneous particle velocity is constant, since when the sound pressure reaches a maximum the particle displacement is a maximum and the particle velocity tends to zero; conversely, when the particle displacement is zero, the velocity is at a maximum. Thus p/v = constant. This constant can be shown to be equal to the product of the speed of sound, c, and the density of the medium, ρ. It is called the *characteristic impedance* of the medium, symbol Z:

$$Z = \rho c \qquad [1.6]$$

The mean sound energy per unit volume of the medium (in this case, the air) is called the *sound energy density*, symbol E, measured in J/m^3. It is related to the amplitude of the maximum particle displacement, ξ, the angular frequency of the sound, ω, and the density of the medium, ρ :

$$E = \tfrac{1}{2}(\rho \xi^2 \omega^2) \qquad [1.7]$$

The rate at which energy flows through the medium depends on the sound energy density and on the velocity of the sound wave in the medium. It is called the *sound intensity*, symbol I, measured in W/m^2. At a point in a unidirectional sound energy flow, sound intensity is defined as the sound power through a small area about that point and normal to the direction of propagation, divided by that small area (see

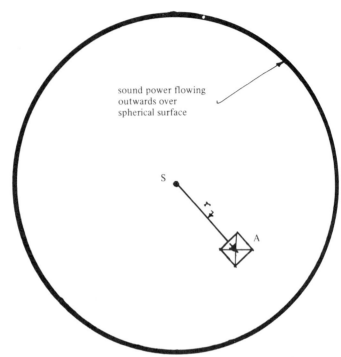

Fig. 1.3. Sound intensity, I, the sound power through unit area normal to the direction of propagation; S = source of sound with a sound power = W, watt; A = unit area of surface normal to propagation direction, m²; r = distance between S and A, m. For a point source of sound in a free field I = W/(4πr²), W/m².

Fig. 1.3). For a point source of sound in a free field, at a distance, r, from the source,

$$I = W/(4\pi r^2) \tag{1.8}$$

Power, W, is equal to energy divided by time, and energy is equal to force times distance, the following relationships can be stated:

Intensity = (Force × Distance)/(Area × Time)
        = (Force/Area) × (Distance/Time)
        = Pressure × Velocity

Sound intensity is a vector quantity, and in practice it may be expressed as the time averaged product of the instantaneous sound pressure, p_i, and the corresponding instantaneous particle velocity, v_i. In a particular direction x:

$$I_x = <p_i v_{i(x)}> \qquad [1.9]$$

where ' $< >$ ' implies a value averaged over time.

It is difficult to measure particle velocity directly, but, using the finite difference approximation it can be derived by integrating over time the difference in sound pressure at two points separated by a distance Δx located along a sound intensity vector:

$$v_{(x)} = \frac{1}{\rho_0} \int_0^t \frac{p_2 - p_1}{\Delta x} \, dt \qquad [1.10]$$

The average sound pressure at the midpoint between the two locations is:

$$p = \frac{(p_1 + p_2)}{2} \qquad [1.11]$$

Thus the vector quantity, sound intensity, may be derived from measurements of sound pressure at two points located close together. However, this type of measurement has only been developed fairly recently and care must be taken to achieve valid results.[1.1]

In practice, the intensity or pressure of the sound is not expressed in arithmetic units. Time-averaging is usually carried out on a root mean square (rms) basis. In addition, because of the very wide range of intensities and pressures to which humans respond, and because increments in intensity are perceived as equal when the ratio between the successive intensity changes is kept constant, logarithmic units, called decibels are used, symbol dB. The reference levels for these ratios have been internationally standardised. For sound intensity the relationship is:

$$L_I = 10 \log_{10} (I/I_0) \qquad [1.12]$$

where

    L_I = sound intensity level, dB re 10^{-12} W/m^2
    I = sound intensity, W/m^2
    I_0 = reference intensity, 10^{-12} W/m^2.

In a freely progressing wave, the relationship between intensity and pressure can be shown to be:

$$I = p^2/Z \tag{1.13}$$

therefore, for sound pressure,

$$L_p = 10 \log_{10}(p^2/p_0^2) \tag{1.14}$$

where

L_p = sound pressure level, dB re 20 µPa
p = rms sound pressure, N/m^2
p_0 = reference pressure, 20 µPa.

A further quantity, sound power level, is defined in a similar manner:

$$L_W = 10 \log_{10}(W/W_0) \tag{1.15}$$

where

L_W = sound power level, dB re 10^{-12}W
W = sound power, Watts
W_0 = reference power, 10^{-12}W

The relationship between intensity and sound power at different distances from a point source in an unbounded medium was given in Equation 1.8. This is known as distance attenuation, or geometrical spreading, and it is equivalent to a reduction of 6 dB sound pressure level for each doubling of distance from the source.

It should be noted that all the relationships shown above between sound power, sound intensity and sound pressure have been determined for the simplest possible sound source in an unbounded medium. In practice, particularly inside buildings, or out-of-doors near large buildings or hills, reflected sound waves cause these simple relationships to change in complicated ways. When a sound wave travelling in air reaches a boundary, such as the ground or a wall of a room some of the energy will be *absorbed* by the new medium and some will be *reflected* back into the air. The ratio of energy absorbed depends on the ratio between the characteristic impedances, Z_1 and Z_2, of the two media. When the two media have similar impedances, most of the sound energy will be transmitted into the new medium, and little will be reflected; conversely, if the two impedances are different, most of the sound energy will be reflected back into the air. As shown in Equation 1.6, the characteristic impedance of a medium is the product of its density and the speed of sound within it; the latter is also related to the density and elasticity of the medium. Thus if the airborne sound wave impinges on a brick wall, the impedance ratio will be large and most of the energy will be reflected. If the airborne sound wave impinges on a porous, low density material, such as glass fibre, most of the energy will be transferred into the new medium.

In practice, in building and planning acoustics, the quantity that is used to describe sound is the *sound pressure level*, in decibels re 20 μPa, (Equation 1.14). When it is required to determine the sound power of the source or to estimate the sound pressure level at some other location, many assumptions regarding the sound source and the sound fields have to be made. These measurement and calculation methods are the subject of a number of acoustics standards, and they frequently require the use of sophisticated measuring equipment and specially designed very reflective rooms, known as *reverberation chambers*, or highly sound absorbing spaces known as *anechoic rooms*. In the future, as instruments for the measurement of sound intensity continue to be developed, this quantity promises to be very useful for determining the acoustical characteristics of sound sources and sound transmission paths in buildings.

1.3 PERCEPTION OF SOUND BY PEOPLE

An understanding of the perception of sound by people is necessary in the context of acoustics and the built environment. This includes the physiological and sensory-neural characteristics of the human hearing system as well as psychoacoustic effects. The ear is considered to consist of three main sections, the outer ear, the middle ear and the inner ear, or cochlea (see Fig. 1.4). The outer ear has two main parts, the visible outer ear and the ear canal. The *pinna* is the name given to the broad upper part of the outer ear and it is important in enabling people to localize sound direction, particularly in the median plane, which is an imaginary vertical plane through the head, normal to the connecting line between the ears and equidistant from them. If a sound source is in the horizontal plane the sound received by the two ears will differ slightly in phase and intensity and this can be interpreted by the brain to determine its location. However, if the sound source is within the median plane the incident sound waves will be the same for both ears. Directional

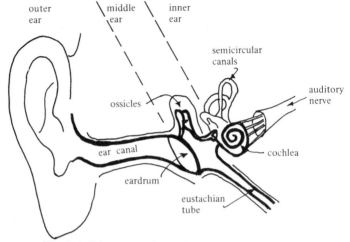

Fig. 1.4. Diagrammatic section of the human ear.

information in this case is apparently deduced by the differential *diffraction* or scattering of the sound at the head and pinna. Diffraction effects are a function of the wavelength of the sound, thus they are different according to frequency. Individuals apparently learn to associate the vertical location of sound sources with the corresponding changes in spectral characteristics caused by this diffraction.[1,2] Correct modelling of pinna shapes has been found to be necessary for research into subjective impressions of concert hall acoustics.

The ear canal is about 25 mm (1 in) long and has a diameter of about 7 mm (1/4 in) although this varies with individuals. The sound waves pass along the canal to the eardrum, or *tympanic membrane*, and set it into vibration. Some of the energy will be reflected back along the ear canal and when the wavelength of the incident energy is integrally related to the length of the canal, reflected waves will be in phase with the incident waves and a so-called *resonance* occurs. This has the effect of increasing (or amplifying) the relative pressure of the sound at that frequency. For typical ear canals, this frequency is about 3 000 Hz.

The middle ear consists of the eardrum, the *ossicles*, which are three small bones situated in an air-filled cavity, and the entrance to the *Eustachian tube*. The eardrum is a flexible membrane which is conical in shape. It is attached to the three small bones which comprise the ossicles, known as the *malleus, incus and stapes* (or hammer, anvil and stirrup) because of their shapes. The ossicles transmit the sound induced vibrations of the eardrum to the *oval window* of the *cochlea*. Because of the relative areas of the eardrum and the oval window, there is a mechanical advantage in this transmission system. The eardrum is only able to respond to the very small changes in air pressure caused by the passage of a sound wave because atmospheric pressure is equalised on both sides of it, through the connection to the nasal cavities provided by the Eustachian tube. (Many people have experienced a temporary 'deafness' when descending in an aircraft, for example; this is because of the rapid change in atmospheric pressure which needs to be equalised in the middle ear.)

The *cochlea*, or inner ear, is the most important part of the auditory system. Various theoretical and physical models have been proposed to describe the manner in which the signals from the vibrating ossicles are transformed into the neural signals which are analysed by the brain and perceived as different sounds. The cochlea consists of a cavity coiled in a flat spiral of two and one-half turns. This cavity is partly divided into an upper and lower gallery by a bony structure and the division is completed by the flexible *basilar membrane*. A second membrane, *Reissner's membrane* also divides the cochlea cavity above the basilar membrane. The cavities are all filled with fluid. The main upper and lower cavities are filled with *perilymph* and the medium one is filled with *endolymph*. (See Fig. 1.5.) The two main cavities are connected by a small opening at the apex of the cochlea called the *helicotrema*. The action of the vibrating stapes on the oval window causes the fluids in the cochlea to vibrate, moving the cochlear partitions with them, as there is not sufficient area in the helicotrema opening to allow free movement of the fluid through it. The pressure is relieved by movement of the *round window* membrane.

Bekesy[1.3] carried out much pioneering work concerning hearing; he found that sound energy causes waves to travel along the basilar membrane from the oval window (attached to the stapes) to the apex, or helicotrema. The location of the maximum displacement of the membrane differs according to the frequency of the sound. For example, low frequency sounds cause the maximum displacement to occur near the apex, although a broad pattern of stimulation occurs; higher frequency sounds cause the maximum displacement to occur near the stapes and they cause little movement of the membrane further along. Naturally, the greater the energy in the incident sound, the greater the amount of the displacement of the membrane at any location.

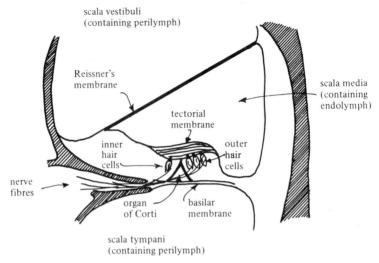

scala vestibuli
(containing perilymph)

Reissner's
membrane

scala media
(containing
endolymph)

tectorial
membrane

inner
hair
cells

outer
hair
cells

nerve
fibres

organ
of Corti

basilar
membrane

scala tympani
(containing perilymph)

Fig. 1.5. Diagrammatic section of the cochlea, showing basilar membrane.

The basilar membrane varies in size and stiffness throughout its length, it is about 35 mm (1.4 in) long overall, and it supports some very complex structures, including the nerve endings of the hearing organ. The actual sensory cells are supported by the basilar membrane; they are called *hair cells* as hair-like projections protrude from their upper ends. There are about 20 000 of so-called Outer Hair Cells (OHC) in three rows and about 3 500 Inner Hair Cells (IHC) in one row. It appears that at least some of the hairs of the OHC span to the underside of the *tectorial membrane*. (See Fig. 1.5). There is a small electrical potential difference between the perilymph in the upper and lower cavities and the endolymph in the middle one, and normally a current leakage occurs through the hair cells. As the cochlear partition is displaced vertically by the sound-induced motion of the fluid, the tectorial membrane is displaced sideways, causing the hair cells to bend. This is thought to release a 'neurotransmitter' which in turn stimulates the acoustic nerve endings. Although there are about 30 000 nerve fibres in each ear, the connection between fibre and hair cell is not on a one-to-one basis. Nerve fibres are either *afferent* or *efferent*: the former carry impulses towards the brain and the latter carry impulses from the brain or nerve centre to the hair cells. Most of the primary afferent nerve fibres (about 95%) innervate the IHCs, with several (about 20) contacting each; in the case of the

OHC, however, several cells contact a single primary afferent nerve fibre, which may extend over half a turn of the cochlea. Efferent fibres enervate both IHC and OHC. It is thought that there are complex interaction stimuli between the cells and this enables the number, location and rate of stimulation of the nerve fibres of the two ears to be interpreted by the brain in terms of frequency, intensity and location of the sound. (It should be noted that this is a simplified explanation of transduction of mechanical energy to electrical energy; further work using advanced electron microscopy promises to provide more detailed descriptions of the cochlear systems.[1.4])

The human audio frequency range is commonly taken to extend from 20 to 20 000 Hertz (Hz) (formerly known as cycles per second) for a normal young person. As people age they frequently lose their high frequency hearing acuity, thus their audio frequency range is reduced. There is some contention as to whether or not this is due solely to the normal ageing process (called *presbycusis*) or whether it is partly the result of exposure to the many loud noises associated with general living styles in technologically developed societies. Kryter[1.5] has argued that this is an important factor which contributes to the loss of high frequency acuity with age: he calls the phenomenon *sociocusis*. The physical result of excessive exposure to noise, whether due to many years of exposure in an industrial working environment, or occurring as a result of acoustic trauma from an explosion, etc. is that the hair cells in the cochlea are irreversibly damaged. Thus they are no longer able to transduce the acoustic energy into neural impulses that can be processed by the brain.

The audio intensity range is from about 0 dB to 120 dB. The lower level is called the *threshold of hearing*, or the minimum level of sound that can be heard under ideal conditions, and the upper level is called the *threshold of pain*, at which the sound is felt rather than heard. However, people do not perceive sounds of all frequencies equally well; they are most sensitive to sound around 3 000 Hz and least sensitive towards the extremes of the audio range.

Many measurements of the average response of people to sound of different frequencies and intensities were made early in the 20th Century, with the advent of radio and telephone communication. People were asked to indicate when a single frequency tone (a sinusoid) sounded equally loud to a reference tone of 1000 Hz at a particular sound pressure level; the graphical results of these experiments are called the *Equal Loudness Contours*.[1.6] Two sounds on the same contour line are perceived as equally loud, and are said to have a *loudness level* of X *phons*, where X is the sound pressure level of the 1000 Hz reference tone, in decibels. The shape of the contours varied considerably according to the level of the reference tone, showing most variation in sensitivity for sounds near the threshold of hearing and becoming almost uniform at very high levels. At about the same time as this research was being carried out, instruments for the measurement of sound, such as *sound level* meters, were being developed, and it was apparent that some recognition should be given to the varying sensitivity of people to sound of different frequencies at different sound pressure levels. Three 'weighting networks' were standardised, the 'A-weighting', corresponding to the 40-phon contour (reference 40 dB at 1000 Hz), the 'B-weighting', corresponding to the 70-phon contour, and the 'C-weighting', which was approximately linear. (See Fig. 1.6 a).) It was originally intended that the different weighting networks should be used according to the magnitude of the sound pressure levels being measured, but this

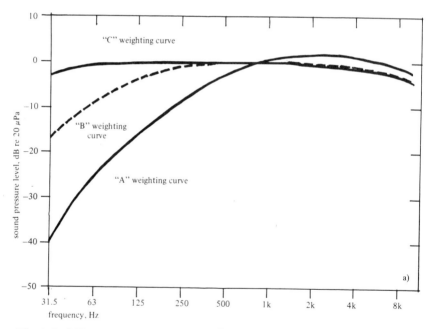

Fig. 1.6. a) Response curves corresponding to the "A-", "B-", "C-" weightings of a sound level meter; b) Equal loudness countours for octave-bands of noise in a diffuse field (after Robinson & Whittle[1.7]); corresponding A-weighting curves shown for comparison.

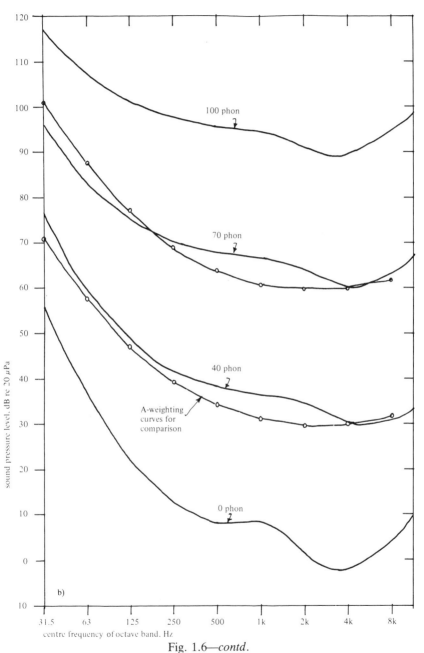

Fig. 1.6—*contd.*

did not prove to be satisfactory in practice. Later investigations by Robinson and Whittle[1.7] found that if the loudness of bands of noise was measured, rather than the loudness of pure tones, the weighting curves were approximately parallel, for any loudness level. The shape of these equal loudness contours was similar to the original 40-phon contour for pure tones, i.e. the contour incorporated in the A-weighting network. Since most sounds that are of interest are more 'noise'-like than pure-tones, or sinusoids, the A-weighting is now normally used whenever measurements are made to represent human response to sound (see Fig. 1.6 b)).

It should be mentioned that the use of the A-weighting network to measure the loudness of different sounds is indeed a simplification; there are more sophisticated methods available for calculating the loudness of complex sounds. In particular, Zwicker[1.8, 1.9] and his colleagues have continued to investigate this problem and they have designed a special Loudness Meter which they claim gives better correlation with human perception of the loudness of complex sounds.

Although 20 Hz has normally been considered as the lower limit of the audio range, with the advent of powerful low frequency sources, such as rockets and very large machines and combustion devices, it is apparent that lower frequencies can be heard. Some experiments have been conducted to determine hearing thresholds at frequencies as low as 2 Hz, but there is still uncertainty on this topic.[1.10] Some experiments have indicated that the dynamic range of hearing is much less at very low frequencies, thus a sound may change from being barely audible to extremely loud with a relatively small change in sound pressure level.

1.4 VIBRATION PERCEPTION

Whereas the ear is the specialised receptor for sound in the audio range, many parts of the body may respond to vibration. At frequencies below about 15 Hz the main receptor is the non-auditory labyrinth, situated adjacent to the cochlea. Above this frequency the main response is through the skin. The main vibration frequency range of interest for human perception is from about 0.5 Hz to 100 Hz. As is the case with audible sound, response to vibration of a given magnitude depends on its frequency; in addition, the direction of the movement is important, as is the position of the human receptor, i.e. whether a person is standing, sitting, or lying down. A co-ordinate system is used to refer to the

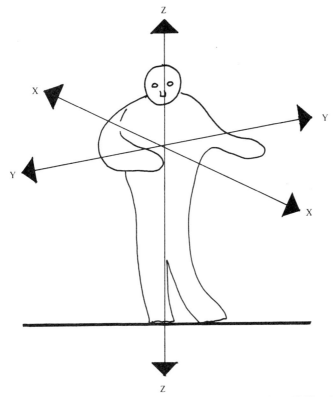

Fig. 1.7. Co-ordinate system for human response to vibration: X-X = fore/aft; Y-Y = side/side; Z-Z = up/down.

direction of the vibration with respect to the human body: 'x–x' refers to fore and aft movement, 'y–y' to sideways movement and 'z–z' to up and down movement (see Fig. 1.7). The magnitude of vibration may be measured either as displacement, velocity or acceleration, but the latter is the quantity usually measured, using an *accelerometer*. A considerable amount of research concerning human perception of vibration has been carried out, but, until recently, it was primarily concerned with comfort of ride in transportation vehicles, or with the ability of pilots to function under conditions of severe vibration, etc. International and National Standards give guidance to the evaluation of the effects of vibration on people.[1.11, 1.12] Three sets of recommended maximum exposures are used: the first relates to the preservation of health or safety (the safe exposure limit), which is set at approximately half the threshold of pain. The second relates to the preservation of working efficiency (fatigue-decreased proficiency boundary), beyond which there is a significant risk of impaired working efficiency: the limits are one-half of those for maximum safe exposure. The third set of exposures are for the preservation of comfort (reduced comfort boundary), which relate to preventing difficulties in writing, drinking, etc.[1.13] In each case the recommended maximum levels depend on the frequency of the stimulus and on the duration of exposure. For example, the maximum human sensitivity to vibration in the z-axis is in the range of 4 to 8 Hz, where for the preservation of comfort a maximum acceleration of 0.1 m/s^2 is the recommended limit for an exposure duration of 8 hours; for the x- and y-axes the maximum sensitivity lies between 1 and 2 Hz over which range, for the preservation of comfort, a maximum acceleration of about 0.07 m/s^2 is the recommended limit for an 8-hour exposure duration. In buildings it is recommended that allowable vibration levels be set close to the threshold of perception.

In some circumstances it is necessary to set building vibration limits to protect sensitive equipment and instrumentation, for example in microelectronics manufacturing establishments. It has been suggested that levels a factor of 5 below the human threshold of detectability may be required in such cases.[1.14, 1.15]

1.5 PSYCHOACOUSTICS

This topic covers a wide range of studies, including laboratory investigations of masking, pitch perception, the loudness of complex sounds, localisation of sources, and both laboratory and field investigations of the reactions of people to noise, including annoyance caused by sleep and activity interference, requirements for speech communication, and subjective impressions of concert halls.

Masking is defined as the number of decibels a sound has to be raised, compared to its perceptible level when heard alone, when in the presence of another sound. Although the issue is complex, depending on the relative frequencies of the signals, information content, etc. generally a low frequency sound masks a higher frequency one more than vice versa. The temporal relationship between two signals is also important. For example, in *forward masking* a second signal may not be perceived if it occurs too soon after the first one. In *backward masking*, the second signal may inhibit interpretation of the earlier signal. Masking has positive and negative effects. For example, if there is a high background sound level in a classroom, the lecturer's voice may not be intelligible to the students. On the other hand, if there is a steady hum of an air-conditioner, conversation in the room next door may be reduced from being distracting to being inaudible.

Localisation and *temporal resolution* of sound is important in auditorium design, particularly if electroacoustics is used. For realism it is necessary that the perceived visual and aural sources are at the same location. It has been found that the direction from which the first sound wave received has come determines the apparent location of the source, even if the succeeding waves are louder; therefore it is necessary that the path travelled by an acoustically amplified sound is greater than that from the original source to the listener. Direct and reflected sound waves are usually perceived as one signal provided that they arrive within about 50 ms; signals arriving later than this, if they are sufficiently loud, may be perceived as distinct echoes.

Assessment of the acceptability of sounds in the community environment has been the subject of many studies. It appears that there is not a simple dose–response relationship between exposure and an individual's reaction to environmental noise, although if large enough samples are taken in social surveys it is possible to estimate the average reactions of groups of people. A number of non-acoustic factors must be taken into account, such as personality characteristics, life-styles, general liking or disliking of a neighbourhood, etc. Specific noise sources such as aircraft may arouse feelings of fear, which will affect annoyance responses. Any characteristics which make a noise source more noticeable, i.e. if it has a strong tonal component such as a hum or whine, or if it is impulsive, will tend to influence its acceptability.

1.6 SOUND SOURCES

1.6.1 Frequency Analysis

It will be evident from the foregoing discussion that perception of any sound will be dependent on its characteristics. In most cases, the overall A-weighted sound pressure level, in dB re 20 mPa, gives a good correlation with overall loudness. However, the overall dB(A) level does not indicate the distribution of sound energy in low, medium and high frequency components. This information is necessary for acoustic design calculations, since the large differences in wavelength over the audio frequency range mean that control measures have different degrees of effectiveness according to the frequency of the sound being considered. *Frequency analyses* are therefore carried out in order to determine the *sound spectra*. Although the unit of frequency, the Hertz, is linear, human perception of changes in pitch is similar to the perception of changes in sound pressure; i.e. if the ratio between two successive pitch differences is constant, the pitch interval sounds constant. Thus a change of frequency from 250 to 500 Hz sounds like the same extent of pitch change (or musical interval) as a change of frequency from 500 to 1000 Hz. (This is not strictly true in musical terms, as there are subtle differences in changes of musical pitch versus frequency; nevertheless, for acoustical purposes the relationship is sufficiently accurate.) In order to take into account this logarithmic perception of frequency changes, a logarithmic scale for frequency is normally used.

Frequency analysis consists of passing the signal through either one-octave or one-third octave band filters. An octave is the frequency interval between f_1 and $2f_1$ and it is described by its geometrical centre frequency, e.g. the octave band centred on 1000 Hz ranges from 707 to 1414 Hz. The three one-third octave bands within this octave have centre frequencies of 800, 1000 and 1 250 Hz respectively (see Fig. 1.8). If the signal has equal energy per Hertz, the one-third octave band levels will be 5 decibels lower than the corresponding octave band level. Two commonly used test signals are also shown on this figure. 'White noise', which is analogous to 'white light', has equal energy at every frequency; thus, when grouped into octave bandwidths there is a 3 dB increase in level per octave. 'Pink noise' has equal energy in each octave band. It should be noted that sound pressure level, in decibels, which is a logarithmic unit, is plotted on a linear scale, whilst frequency, in Hertz, which is a linear unit, is plotted on a logarithmic scale for the reasons outlined above.

Occasionally, particularly for machine noise diagnostic purposes, one- or one-third octave bandwidths are too coarse and narrower bandwidths are necessary. The band-spectrum levels of a sound source may be plotted as sound pressure levels or as A-weighted sound pressure levels— the latter are more useful if it is required to identify the loudest components. In Fig. 1.9 a typical spectrum of urban road traffic is shown both as unweighted sound pressure level and as A-weighted sound pressure level. It can be seen from the A-weighted curve that in this case the loudest components, which must be reduced if the overall sound

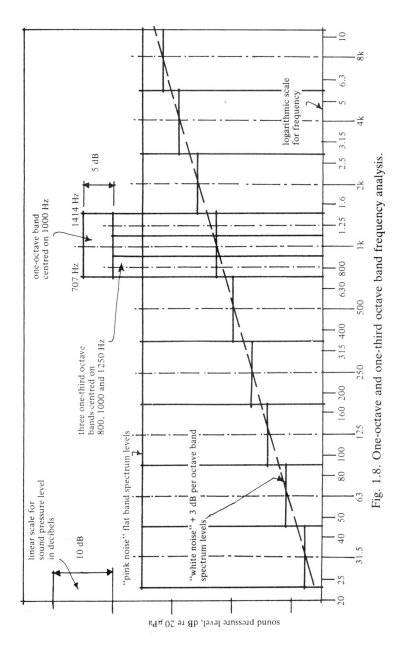

Fig. 1.8. One-octave and one-third octave band frequency analysis.

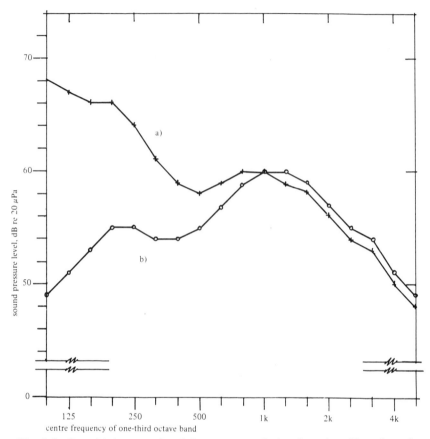

Fig. 1.9. One-third octave band frequency analysis of road traffic noise; a)
unweighted, or 'Linear', b) 'A-'weighted.

level is to be lowered, are around 1000 Hz and above, not at the low
frequencies which are prominent in the Linear curve.

1.6.2 Computations with Decibels

If it is required to reduce the received noise levels, for example, the
one- or one-third octave band characteristics of the control device
are applied individually to the relevant source noise spectrum. The
attenuated levels are then recombined to obtain the overall A-weighted
noise level. Although such computations must take into account that the

TABLE 1.1
Combination of Decibels

Difference between L_1 and L_2, dB	Add to L_1, dB
0, 1	3
2, 3	2
4, 5, 6, 7, 8, 9	1
10, >10	0

decibel is a logarithmic quantity, in most cases a simple table may be used for combination purposes (see Table 1.1).

For example, if a machine has the octave-band spectrum shown below and it is intended to separate it from an office by building a wall with the sound transmission loss characteristics given, what will be the reduction in overall A-weighted sound pressure level? (Note, this example is simplified and does not consider all factors that must be taken into account in practice.) The calculation is shown in Table 1.2.

In order to determine the overall A-weighted transmitted sound level, the octave-band spectrum levels are combined. It is preferable to place the levels in order, from lowest to highest, and then to combine them as shown in Table 1.3.

The original machine sound level in this example can be computed in the same way: it is 81 dB(A), therefore the attenuation provided by the wall in this example is (81–58) = 33 dB(A).

1.6.3 Sound Directivity

Only the simplest type of sound source has been considered up to now in the discussion of sound generation and propagation. The concept of the small, point source of sound in an unbounded medium is useful, but in practice, many sources are not small, they may radiate more sound in some directions than in others and they frequently operate in an enclosed space, or at least close to the ground or to large reflecting surfaces out of doors. If the sound pressure level is measured at a fixed distance around a source it is usual to find that the level varies according to the location of the microphone. This is quantified as the *directivity factor*, symbol Q. It may be measured if a source of power W, watts, radiates energy into an unbounded medium (in practice, a highly absorbing room known as an anechoic room). The sound pressure at a particular distance and direction is compared with that measured at the same

TABLE 1.2
Estimation of Noise Reduction

	Centre frequency of octave band, Hz					
	125	*250*	*500*	*1 000*	*2 000*	*4 000*
Machine noise levels, dB re 20 μPa	88	84	77	76	72	68
Apply A-weighting, dB	−16	−9	−3	0	+1	+1
A-weighted noise levels, dB(A)	72	75	74	76	73	69
Deduct sound transmission loss of wall, dB	18	22	27	32	29	31
Transmitted sound level, dB(A)	54	53	47	44	44	38

TABLE 1.3
Example of Combining Band Spectrum Levels

Rearranged octave band sound levels, dB(A)	38	44	44	47	53	54
L_1-L_2		6	1	1	2	1
Add to L_1[a]		1	3	3	2	3
Progressive total, dB		45	48	51	55	58
Therefore, overall level is 58 dB(A)						

[a] L_1 is always the higher of the two values being compared.

distance and direction for a non-directional source radiating the same sound power. It is found from the following:

$$Q_\theta = p_\theta^2/p_s^2 \qquad [1.16]$$

where

Q_θ = the directivity factor in direction θ,
p_θ^2 = the sound pressure measured at a distance r from the source in the direction θ,
p_s^2 = the sound pressure measured at the same distance r from a non-directional source, both sources radiating the same sound power, W.

Directivity may also be expressed in a logarithmic form, where the *directivity index*, symbol *DI*, in a direction θ, is defined as

$$DI_\theta = 10 \log Q_\theta \qquad [1.17]$$

or

$$DL_\theta = L_{p\theta} - L_{ps} \qquad [1.18]$$

where

$L_{p\theta}$ = the sound pressure level, in dB re 20μPa,
measured in the direction θ and at distance r
from the directional source
L_{ps} = the sound pressure level, in dB re 20μPa at distance r
from the non-directional source.

Although the directivity index may be measured as the overall A-weighted sound pressure level, it will usually be found to vary according to frequency. Most sources tend to become more directional in character as the radiated frequency increases, and at high frequencies the signal may form a highly directional 'beam'.

Another important effect associated with the type of source and its directivity is its efficiency as a sound radiator. The most efficient radiators are non-directional sources; they are known as *monopoles*. In addition to point sources, large plane surfaces such as walls and floors radiate sound in a similar, efficient manner. If a source emits sound, not by changing its volume, as does a point source, but by changing the location of its centre of gravity, it is known as a *dipole*. Since much of the acoustic energy is dissipated as the fluid around the body is shifted backwards and forwards, this type of source is a less efficient sound radiator. The radiation pattern is a 'figure of eight', i.e. much more energy is radiated in the direction of oscillation and much less normal to this direction. Other sources emit sound due to their change of shape, rather than a change of volume or location; these are known as *quadrupoles* and these are the least efficient sound radiators. Aerodynamic sources have quadrupole characteristics.

The actual sound radiation patterns are also considerably affected by the space in which the source is located. For example, if a source is close to a large surface, its radiation pattern will be affected. At some distance from a source, a receiver will perceive reflected sounds as well as the direct sound. One of the main problems in acoustics is to predict what the overall effect will be.

1.7 SUMMARY

In this chapter a brief introduction has been given to the physics of sound. The ear and human hearing and reaction to sound have also been briefly discussed. Finally, methods of manipulating decibels have been shown. In the following chapters more detailed information will be given on some of the relevant aspects of acoustics and their application in the built environment.

REFERENCES

1.1. Crocker, M.J. Direct measurement of sound intensity and practical applications in noise control engineering. *Internoise* 84, Honolulu, pp 21-35
1.2. Schroeder, M.J. Models of hearing. *Proc. I.E.E.E.* 63, 1975, pp 1332-1350.
1.3. Bekesy, G.von, *Experiments in Hearing*, McGraw Hill, New York, 1960.
1.4. Special Issue on Hearing Research. *J.Acoust.Soc.Amer.* 78, 1985 (1, Pt 2), pp 295-388.
1.5. Kryter, K.D. *The Effects of Noise on Man*, 2nd Ed. Academic, New York, 1985, Ch. 6.
1.6. Fletcher, H. & Munson, W.A. Loudness, its definition, measurement and calculation. *J.Acoust.Soc.Amer.* 5, 1933, pp 82-108
1.7. Robinson, D.W. & Whittle, L.S. The loudness of octave bands of noise. *Acustica*, 14, 1964, pp 24-35
1.8. Zwicker, E., Deuter, K. & Fastl, W. Loudness meters based on ISO 532B with large dynamic range. *Internoise* 85, Munich, pp 1119-1122.
1.9. Zwicker, E. & Fastl, W. Examples for the use of loudness: transmission loss and addition of noise sources. *Internoise* 86, Cambridge, USA, pp 861-866
1.10. Yeowart, N.S. Thresholds of hearing and loudness for very low frequencies. Ch.3 in *Infrasound and Low Frequency Vibration*, W.Tempest, ed. Academic, London, 1976
1.11.—*Guide for the Evaluation of Human Exposure to Whole-body Vibration*, ISO 2631-1978, International Standards Organisation
1.12.—*Vibration and Shock—Guide to the Evaluation of Human Exposure to Whole-Body Vibration*, AS 2670-1983, Standards Association of Australia
1.13. Griffin, M.J. Effects of vibration on humans. *Internoise* 83, Edinburgh, pp 1-14
1.14. Gordon, C.G. & Ungar, E.E. Vibration as a parameter in the design of microelectronic facilities. *Internoise* 83, Edinburgh, pp 483-486
1.15. Ungar, E.E. & Gordon, C.G. Vibration criteria for microelectronics manufacturing equipment. *Internoise* 83, Edinburgh, pp 487-490

CHAPTER 2

Noise in the Community

2.1 INTRODUCTION

Community noise is often considered to be a form of environmental pollution, and as such regulations intended for its control may be the responsibility of the same authority that is concerned with air and water pollution. However, it must be emphasized that there are some significant differences in the propagation of noise compared to the spread of air or water pollution and it is important that this is recognized. One aspect that noise has in common with other pollutants, however, is that the most effective controls are those effected at the source, together with the application of careful regional and local land use planning measures. Community noise levels are not usually sufficiently high to cause permanent loss of hearing (see Section 1.3); however, they may cause considerable annoyance, and, in some cases this can lead to stress and deleterious health effects (see Section 1.5). In this chapter various types of community noise source will be discussed, together with methods of measuring and assessing the acceptability of community noise levels. Since in many cases the source of the noise may be at a considerable distance from the recipients, the effects of topography and meteorology on sound propagation must also be considered.

2.2 COMMUNITY NOISE SURVEYS

There have been many surveys conducted in many countries which have attempted to determine the relationship between noise exposure and the annoyance caused to a community. Annoyance is a difficult property to measure objectively—it is related to, but not identical with, loudness. Two sounds may be equally loud, but not equally annoying, and any one sound may or may not be acceptable according to circumstances, such as the time at which it is heard. Annoyance is clearly related to the disturbance caused, and this in turn depends not only on the level and spectrum of the interfering noise, but on the involvement of the disturbed person in a particular activity (or lack of activity in the case of sleep). The principles of masking and appropriateness largely determine the acceptability of community noise. If a noise cannot be heard, it will not be annoying—this means that if the background noise levels in an area are generally moderate to high, many individual noises, that might otherwise be annoying, will not be noticed. However, community noise levels are usually lower at night, and individual noises may then intrude above the background and cause problems. Appropriateness refers to the general character of a neighbourhood—if it is predominantly industrial then machinery noise is expected; however, the same noise emitted in a suburban residential environment may lead to complaints.

Community noise surveys consist of two parts—assessment of the actual noise levels in an area, usually by physical measurement, and determination of peoples' reactions to the noise, by completion of questionnaires. Considerable care is needed in the design and administration of the questionnaire used to examine people's attitudes to community noise (which is true for any type of social survey). Judging by the results of many community noise surveys, it appears that however carefully the research programme is designed and carried out, the correlation between noise exposure and individual subjective annoyance reactions is poor—typically the correlation coefficients are about 0.3 to 0.4, which implies that factors other than noise are the main determinants of people's responses. In order to obtain better correlation between the noise levels in an area and annoyance reactions, it is necessary to pool the responses of groups of people. It is evident that there are large differences in individual tolerance to noise in the community, and people who are generally satisfied with the living conditions in a neighbourhood, e.g. with access to transportation, shopping facilities, etc. will show more tolerance towards noise. Conversely, 'excessive noise' is sometimes used as a complaint when it is other factors that are really causing annoyance. There is some evidence to show that there may be about 10% of any population who are not at all concerned about noise, and another 10% that will continue to express annoyance, even at very low levels of exposure. Schultz commented that when the noise exposure in a neighbourhood is high, the response of people is more uniform and non-acoustical variables play a smaller role.[2.1] From an analysis of a number of social surveys carried out in different countries (many of which used different nomenclature in their questionnaires) he proposed that the people comprising the upper 29% or so of responses on annoyance scales could be considered to be 'highly annoyed'. He therefore suggested that noise limits could then be based on the 'acceptable' percentage in a community who would be expected to be highly annoyed. If no one at all were to be highly annoyed, the limit would be of the order of a little over 40 dB(A) daytime and 30 dB(A) night-time. If, say 10% of people highly annoyed was accepted as a reasonable compromise, these limits could be about 20 dB(A) higher.

The other aspect of research studies to determine the response of people to noise which needs to be resolved is the way in which their noise exposure should be measured and described. Typically, the noise levels are measured at locations considered to be representative in each neighbourhood. For practical reasons, the microphones are usually located outside buildings (to avoid interference from noise generated inside). However, there may be little or no correlation between the measured external noise levels and the actual noise levels, originating from outside, to which a person inside the building is exposed. Intervening factors are whether or not the occupied room(s) are exposed to the main external noise source, whether or not windows or doors are open to provide natural ventilation, whether or not there are significant noise levels generated indoors (radio, TV, machinery and appliance noise), and the time(s) of day that the building is occupied, compared to the operating time(s) of the main external noise sources.

In many situations, the external noise levels vary considerably according to the time of day, day of the week, and sometimes with the change of season. If it is required to determine the complete noise exposure environment, it would be necessary to monitor the noise levels continuously. In some cases, this is done, but for most practical situations the amount of data generated is excessive. Some type of time-sampling procedure is normally adopted. One method used is to take sample measurements for a short time at regular periods, e.g. for two minutes in each hour over a 24- or 48-hour period. There have been several studies concerning the errors introduced by sampling, and it has been found that both the noise level distribution and the descriptor used affect the results. One of the earliest attempts to describe time-varying noise, particularly from road traffic, was to determine the noise level exceeded for a certain percentage of the time period. The level exceeded for 10% of the measuring time period, designated L_{10}, was chosen in an early London noise survey, to represent the higher levels of environmental noise.[2.2] The levels exceeded for 90% or 95% of the time, L_{90} or L_{95}, is frequently chosen to represent the lower levels, or the background sound level in an area. These percentile values are towards the extremes of the sound level distribution, and, for a given level of statistical accuracy, need longer measurement time periods. Schultz found that for periods of low traffic flow, a two minute sample had a low probability of giving the result within 2 dB of the correct one. If the level exceeded for 50% of the time, L_{50}, is chosen as the descriptor, accuracy is improved. Another way of overcoming the problem of excessive data collection is to use 'microsampling', that is recording ten seconds every 5 minutes to give a total cumulative sample of 2 minutes every hour.[2.3]

Some of these time-sampling problems have been overcome with the advent of more sophisticated instrumentation such as 'black boxes' which can be left, unattended, to monitor noise levels continuously, and which will automatically calculate the percentile sound levels. The disadvantage of such instruments is that there is no analogue record of the sound that has been measured, and thus any unusual occurrence that may have affected the measurements is undetected; it is also not usually possible to obtain a frequency analysis, only the overall linear or A-weighted sound levels are provided.

One of the most frequently-used descriptors for time-varying environmental noise now is the equivalent energy level, L_{eq}, more properly defined as the 'Equivalent continuous A-weighted sound pressure level, $L_{Aeq, T}$', dB(A). This is the value of the A-weighted sound pressure level of a continuous steady sound that, within a measurement time interval T, has the same mean square sound pressure as a sound under consideration whose level varies with time.[2.4, 2.5] The physical description is:

$$L_{Aeq, T} = 10 \log \left(\frac{1}{t_2 - t_1} \int_{t_1}^{t_2} \frac{p_A^2{}_{(t)}}{p_0^2} \, dt \right) \qquad [2.1]$$

where

t_1, t_2 are the start and finish times of the measuring period T,

p_0 is the reference sound pressure, 20 μPa,

$p_{A(t)}$ is the instantaneous A-weighted sound pressure at time t

Many acoustic instruments are available to obtain this quantity directly, including automatic monitors as mentioned above. The equivalent energy level has the advantage that all the sound energy received during the measuring time period is included, and unlike percentile levels, no assumptions have to be made about the sound level distribution over time.

2.3 COMMUNITY NOISE SOURCES

There are several broad categories of community noise source; they include transportation vehicles (road, air, rail and water); industry; outdoor sport and recreation; and 'people noise' including domestic appliances and pets. Although there are particular situations where aircraft noise or industrial noise predominate, in most industrialized countries it has been found that the most prevalent source of community noise is road traffic.

2.3.1 Road Traffic Noise

One of the first major noise surveys to be carried out was in the central area of London in the 1960s. This resulted in the finding that at 85% of the 540 measuring sites (at locations that were chosen simply as the intersection points of regular grid lines on a map of the area) road traffic noise was most frequently heard. It also more frequently produced the highest level of noise than any other source.[2.2] This finding was replicated in a number of later studies in many parts of the world, and it has lead to a great deal of research concerning the noise emission characteristics of vehicles, noise emission from traffic streams and motor vehicle and road traffic noise reduction. It is unfortunately true to say that although there has been some progress, particularly in the reduction of noise from large commercial vehicles, it will still be many years before there is a general reduction of traffic noise in most communities.

Although traffic noise is the result of emissions from many individual vehicles, the noise is typically perceived as coming from the traffic stream. There are several methods used to predict the resulting noise levels, some of which have been based on theoretical considerations of traffic behaviour, and others which have been determined on an empirical basis. One of the first matters to be decided is the *acoustic descriptor* to be used to characterize the constantly varying sound levels that occur as vehicles pass by a particular receiving location. As mentioned above, in Britain, as a result of the early London noise surveys, the descriptor L_{A10} was decided upon as representative of the noisier events. However, it does not provide any information regarding the range of noise levels experienced; for example, with a low traffic flow rate the noise level may frequently drop down to the background level, and the range may be 30 decibels or more; as the traffic flow rate increases, traffic noise will dominate the noise climate all the time and there will be a much smaller difference between the noisier and quieter periods. Another descriptor suggested, arising from a social survey of occupants of dwellings in Britain in 1967 was the Traffic Noise Index, TNI.[2.6] This takes into account the variability of the noise level, as expressed as the difference between L_{10} and L_{90}, not just the noisier events; however it has the disadvantage that as the traffic flow rate increases, and L_{90} increases, the TNI becomes smaller.

An alternative, the Noise Pollution Level, NPL, which still takes into account the variability in the noise levels, was suggested by Robinson:[2.7]

$$NPL = L_{eq} + k\sigma \qquad [2.2]$$

where

L_{eq} is the A-weighted equivalent sound pressure level
k is a constant ($= 2.56$)
σ is the standard deviation of the instantaneous sound pressure level

However, although NPL in particular is conceptually attractive, the general low correlation found between *any* noise descriptor and individual annoyance reactions puts into doubt the need for a relatively complicated descriptor. Since for some years there has been a growing acceptance of $L_{Aeq, T}$ to describe noise of any origin it is logical to use it for road traffic noise. If it is more appropriate to use L_{10}, because of legislative requirements for example, the close correlation which exists between $L_{Aeq, T}$ and $L_{A10, T}$ is helpful. In most cases where traffic noise is a problem the traffic flow rate is such that $L_{A10} \cong L_{Aeq} + 3$.

Where a new road is proposed, or an existing road is to be modified, it is necessary to predict the noise levels that will be received at nearby sites or buildings. At some distance from a road, each *individual vehicle* may be considered as a 'point' source of sound, and thus the maximum

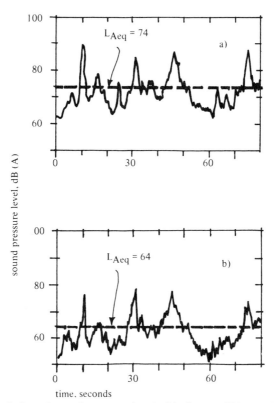

Fig. 2.1. Variation of sound pressure level with time at different distances from a busy road; a) microphone at 9 m (30 ft) from centre of nearside traffic flow, b) microphone at 36 m (120 ft) from centre of nearside traffic flow.

sound level received as a vehicle passes by would be expected to attenuate at about 6 dB for each doubling of distance from the source (see Section 1.2). However, the noise emitted from a *stream of traffic* propagates more like that from a line source, which theoretically attenuates only at a rate of 3 dB for each doubling of distance. Thus, as the receiving point moves further from the line of traffic, the range of traffic noise levels is reduced, as the maxima attenuate at a greater rate than the average. Fig. 2.1. shows typical A-weighted sound pressure level time traces of road traffic noise at different distances from a busy road. Several theoretical and empirical methods for predicting traffic noise levels have been developed over the years: some are quite simple and others attempt to take into consideration many factors, such as the types of vehicles in the traffic stream, vehicle speed, the road surface and gradient, shielding by topography, barriers and buildings, etc. The general form of many of the prediction equations is as follows:

$$L_{Ax} = A + B \log Q + Cp + Dv - E \log d \qquad [2.3]$$

where

L_{Ax} is the chosen traffic noise descriptor, dB(A)
A, B, C, D and E are constants
Q is the total vehicle flow rate, in vehicles per hour
p is the percentage of heavy vehicles
v is the average vehicle speed, in km/h
d is the distance from the centre of the nearside carriageway, m.

There are slight variations in the definition of 'heavy vehicles' in different countries, but the term is generally used to describe commercial vehicles having more than four wheels (i.e. two-axled-vehicles with dual wheels on the rear, or vehicles with more than two axles). Such vehicles, because of their size and construction, usually emit considerably more noise than individual passenger vehicles. This factor is also recognised in higher noise level limits for regulatory purposes.

The vehicle speed factor is included because vehicle noise emission is related to engine speed. For freely flowing traffic there is a close relationship between engine speed and vehicle speed. At high speeds, above about 80 km/h (50 mph), which commonly occur on motorways and freeways, tyre/road interaction also contributes to the emitted noise level. However, in congested urban traffic, there is little correlation between vehicle speed and engine speed, the tyre/road contribution is lower and modifications are necessary to the prediction equation.

As mentioned above, as the distance between the traffic stream and the receiver increases, the noise level will be reduced because of geometrical spreading of sound energy. Although, theoretically, on a multi-lane road, the distance between a given receiver and the vehicle will vary according to the lane on which the latter is travelling, it has been found in practice that unless there is a divided road and the carriageways are separated by more than about 5 m (16 ft), all the vehicles in both directions may be assumed to travel along a line that is called the 'centre of the nearside traffic flow'. In other words, if there is a six-lane divided highway, with a narrow median strip and three lanes in each direction, all lanes being used, 'd' is measured from the centre of the nearside carriageway to the receiving point.

A comprehensive prediction method for noise emitted from road traffic is that used in Britain for determining eligibility for compensation.[2.8] This takes into account ground reflection effects, shielding by other buildings and barriers, road gradients, etc. See Section 3.3.2 for an example of the use of this method.

2.3.2 Aircraft Noise

Although, as stated earlier, road traffic is the most widespread source of community noise, for those people living or working near an airport, aircraft are by far the most prominent sources. It should be remembered that for most of their operational time, aircraft are far from communities and have little effect on them; however, when an aircraft lands or departs from an airport it is usually extremely noisy. If all airports were separated by large distances from noise-sensitive land-occupancies, there would be no aircraft noise problem. However, unfortunately, many existing airports are quite close to the communities which they serve. This has occurred sometimes because of the growth of an old airport which originally catered only for infrequent services of small aircraft, and sometimes because of insufficient planning controls to prevent new communities extending to the vicinity of the airport, attracted there often by the provision of good land transportation corridors.

Since the advent of jet-propelled aircraft in the late 1950s there has been much research applied to the reduction of noise at the source. Aerodynamic noise from the jet itself is one of the main sources, and this was found to be proportional to the 8th power of jet velocity. There was a significant break-through with the introduction of the high by-pass ratio, fan-jet engines which, because they are more fuel efficient, have gradually replaced the straight turbo-jet engines. These engines are inherently quieter because the by-pass air exhausts have much lower jet velocities than the main jet and a smoother transition between the jet and the surrounding air is achieved (See Fig. 2.2). However, it is unlikely that there will be any further major advance in noise reduction in the near future, as other sources, such as the fan, the compressor, the turbine and even airframe noise now contribute significantly to the levels heard on the ground.

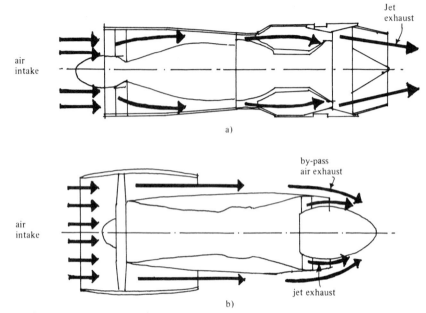

Fig. 2.2. Aircraft engines; a) early axial flow turbo-jet, b) high by-pass ratio engine.

Most countries now have legislation controlling the noise emission from aircraft. Civil aircraft are usually required to conform to the noise certification limits agreed to by the International Civil Aviation Organisation, ICAO. These limits have been made more stringent as noise control technology permitted, and they refer to specific locations on the ground for an aircraft flying under specified operational conditions. Larger, heavier aircraft are permitted higher certification levels than smaller ones. For each aircraft type, a 'noise print' may be plotted showing the extent of noise impact on the ground during take-off and landing. The effectiveness of noise controls may be seen by comparing, for example, the noise print of an early, narrow-bodied jet airplane with that of a more recent, high by-pass ratio, wide-bodied airplane (Fig. 2.3). Both aircraft carry approximately the same number of passengers over the same distance.

For planning purposes it is necessary to know not only the maximum aircraft noise levels to be expected at a particular site, but also the frequency of aircraft operations. There are several methods in use to

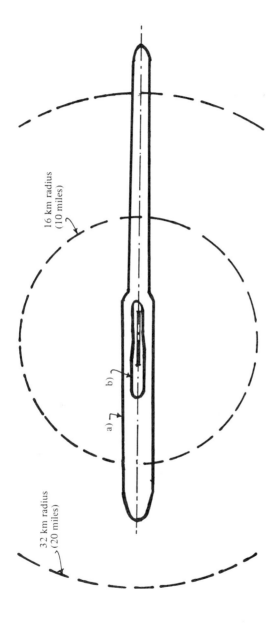

Fig. 2.3. 'Noise-prints' of long-range, 200 passenger aircraft on the ground; a) early, narrow-bodied jet, b) wide-bodied, high by-pass ratio jet.

assess the noise impact of an airport on its surroundings. When jet aircraft were first introduced in the United States social surveys indicated that the effect of the high-frequency jet 'whine' was not adequately described by an overall 'A'-weighted sound pressure level, and a new frequency weighting, with additional high frequency emphasis, known as the Perceived Noise Decibel, PNdB, was proposed, and adopted for New York's Kennedy Airport. However, not everyone agreed on the necessity for this special weighting system.[2.9] The PNdB assessment was further refined to include a duration allowance and a correction for pure tones. This is known as the Effective (tone-corrected) Perceived Noise level, EPNLdB, and requires one-third octave band monitoring of aircraft flyovers. In order to assess the total noise impact of an airport, an estimate is made of the number and type of aircraft movements for each aircraft type at some date in the future. The EPNLdB level is estimated at selected ground positions, for each projected movement, taking into account expected runway usage, aircraft loading and flight path. A penalty is included for night-time operations and the result is known as the Noise Exposure Forecast, NEF, for a particular year.

$$NEF = 10 \log \left[\Sigma_{i,j}(N_{i,j} + 16\,N'_{i,j})\,10^{EPNL_{i,j}/10} \right] - 88 \qquad [2.4]$$

where

$EPNL_{i,j}$ is the energy mean value of EPNL for aircraft of type i performing operation j

$N_{i,j}$ is the number of such aircraft operating in daylight hours, 0700-2200

$N'_{i,j}$ is the number of such aircraft operating at nightime, 2200-0700 hours

$EPNL = PNL_{max} + 10 \log \left[(t_2 - t_1)/T_0 \right]$

$(t_2 - t_1)$ is the time during which the level is within $-10dB$ of the maximum noise level

T_0 is the normalized duration = 15 seconds.

Contours are drawn through positions on the ground with the same NEF value. Guidelines are given for the compatibility of different land-uses with NEF zones. (It should be mentioned that although the NEF system was developed in the United States, many of the airport authorities in that country use other assessment methods). A socio-acoustic investigation to assess aircraft noise impact on Australian

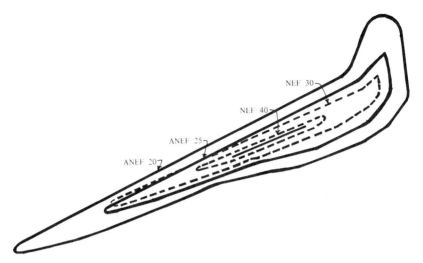

Fig. 2.4. Noise Exposure Forecast, comparison between NEF contours and Australian Noise Exposure Forecast, ANEF, contours and recommended compatibility for residential land-use, for a medium-sized airport. Not compatible NEF > 40 ($\cong$ 2 km^2); ANEF > 25 ($\cong$ 20 km^2); Conditionally Acceptable NEF 30-40 ($\cong$ 10 km^2); ANEF 20-25 ($\cong$ 26 km^2).

communities was reported by Hede and Bullen in 1982.[2.10] As a result of this survey the NEF system has been amended for use in this country. It was found that many people were annoyed with aircraft noise in the evening period, which was not reflected in the original system. Thus 'night-time' is now extended to include the evening period, from 1900–0700 hours and 'daytime' is from 0700–1900 hours. The night-time penalty (which in the NEF system is effectively equal to +12 dB for night-time operations) was reduced to 6 dB. New contours have been determined on this basis, known as the Australian Noise Exposure Forecast contours, ANEF. Again, guidelines for compatible land-uses are provided, which tend to be more stringent than those originally suggested in connection with the NEF contours (partly because of the effect of a more temperate climate on life-styles and building construction). It can be seen from Fig. 2.4 that the method of aircraft noise assessment and the compatibility guidelines have a significant effect on the area of land considered to be unsuitable for housing, for example.

In the United Kingdom, in 1963, the Wilson Committee[2.2] introduced the concept of the Noise and Number Index, NNI, based on studies carried out around London, Heathrow airport:

$$NNI = PNdB + 15 \log N - 80 \text{ (daytime)} \qquad [2.5]$$

where

PNdB = the average peak noise level
  N = the number of flights.

It was recommended that NNI should not exceed 50 to 60 units.

More recently, land-use planning legislation has been introduced in a number of Western European countries. For example, in W.Germany the Air Traffic Noise Act of 1971 established noise protection areas, consisting of two zones with equivalent sound levels of more than 67 dB(A) and 75 dB(A), for all commercial airports and military airfields with jet aircraft operations.[2.11, 2.12] There are restrictions relating to buildings in these areas and grants are made for improving building attenuation in the inner zone.

It is extremely important that planners understand the implications of aircraft noise contours around airports, and that they do not allow unsuitable land zoning to occur. This is particularly critical for buildings which rely on natural ventilation, or which have associated outdoor activities, such as residential buildings and schools. Although it is possible to use sealed, massive building construction and air-conditioning to reduce aircraft noise intrusion to acceptable levels indoors, most people like to have some aural contact with the outside world, and if the building is designed to keep out aircraft noise, all wanted sounds, such as approaching visitors, birds, children playing, etc. are usually rendered inaudible. Close cooperation with the airport authorities is necessary to ascertain likely future developments, such as additional runways, or expected increases in aircraft movements, over the life-span of intended building developments.

Although the individual noise levels emitted by helicopters and general aviation aircraft are generally less than those from large commercial aircraft, their movements are less regulated and they can be a significant source of annoyance to the surrounding community. Gummlich reported on surveys carried out around general aviation airfields in Germany and found approximately 50% of the population considered flight operations a nuisance.[2.13] Such airfields are often located in otherwise quiet, non-urban areas and their maximum usage frequently occurs in the evenings and at weekends, which is the time that more people tend to be at home relaxing. Because of the difficulty of assessing the actual flight patterns and aircraft operations, it may be more appropriate to apply restrictions to the allowable total number of movements and also to the times at which they may occur.

For both major and general aviation airports, some pressure can be exerted on operators by using financial incentives. For example, at some airports landing fees are higher for noisier aircraft; curfew exemptions may be given to new-generation quiet aircraft, etc. In some countries, 'noise charges' are used to fund amelioration measures in nearby noise-sensitive buildings.

2.3.3 Railway Noise

The effect of railway noise on communities has not been studied as extensively as has road traffic and aircraft noise. In many cases, the railways have been in existence for a long period of time, and, with their generally predictable noise levels and train pass-bys, there is some evidence to show that they may not cause as much annoyance as other forms of transportation, such as aircraft. A major survey of people's reactions to railway noise was carried out in Britain in the late 1970s.[2.14] The best correlation was found between annoyance reactions and $L_{Aeq, 24 hours}$. No particular 'acceptable level' was found, although the rate of increase in annoyance with increase in noise level was greater above 40 to 50 dB, $L_{Aeq, 24}$. Freight traffic caused more annoyance than electrified passenger trains, and vibration was the most important non-noise impact. Additional problems arise from railway maintenance operations (which frequently take place at weekends, when general rail traffic is lower). In recent years, the noise and vibration associated with the advent of very high speed trains in a number of countries has caused many problems.

There are several noise sources associated with trains. The locomotive itself may be a significant source, particularly if it is powered with a diesel-electric engine. These engines tend to have strong low-frequency components.[2.15] In conventional-speed trains, wheel-rail interaction is the major noise source for the non-powered wagons. 'Roar' noise occurs when there is small-scale roughness, or corrugations on the rails; impact noise occurs when wheels roll over their own flat spots as well as over joints or discontinuities in the rails. For this reason, welded rail is quieter than bolted rail, and good maintenance is very important. Another annoying noise source is 'squeal', which occurs on curves when a wheel is forced to slip laterally on the rail:

the resulting 'stick–slip' action excites the natural vibration modes of the wheel and the resulting noise level may be over 20 dB higher than from the other sources. This may be avoided in part by using larger radius curves.[2.16] When railways pass over bridges, the whole structure may radiate sound energy if there is inadequate structural isolation. Very high speed trains (travelling at over 200 km/h) present additional noise sources. The earliest extensive high-speed railway installations were in Japan, and strong adverse reactions to the resulting noise and vibration resulted.[2.17] A particular problem in Japan is the generally small separation distance between buildings and railway track; in addition, much of the track is elevated to avoid level crossings. Prior to the introduction of noise controls, levels of 85–90 dB(A) at 25 m(80 ft) from trains travelling at 200 km/h(125 mph) on elevated, ballast-less track have been measured, with typical passby durations of 11 to 15 seconds. Environmental quality standards in the vicinity of high-speed railways in Japan were set at 70 dB(A) for residential areas and 75 dB(A) for commercial areas. As well as wheel-rail noise, other sources of noise from high-speed railways include the power collecting devices, aerodynamic noise and structure radiated noise. In Germany, it has been found that at speeds over about 250 km/h (150 mph), aerodynamic noise may be the main problem, particularly from the pantograph.[2.18] This has implications for noise control techniques, since the location of this source is much higher above the track than the wheel-rail source, rendering noise attenuating devices, such as barriers, less effective. (See Section 3.3.3).

There are fewer prediction methods for railway noise than for road traffic. However, the engine noise from a locomotive may be considered as a point source and the wheel/rail interaction noise from carriages and wagons as a line of incoherent dipoles.[2.19] For an individual train hauled by a locomotive, the passby sound pressure level, L_{Ax} may be estimated from the following: [2.20]

$$L_{Ax} = 10 \log [10^{L_{Ax1}/10} + 10^{L_{Ax2}/10}] \qquad [2.6]$$

where:

$L_{Ax1} = L_{Amax1} + 10 \log (d/v) + 8.6$
$L_{Ax2} = L_{Amax2} + 10 \log (L_t/v) - 10 \log [(4D/4D^2 +1)+2\tan^{-1} (1/2D)]+10.5$
$L_{Amax1} =$ maximum A-weighted spl from the locomotive at the reception point

L_{Amax2} = maximum A-weighted spl due to the wheel/rail interaction at the reception point

d = perpendicular distance, track to observer, m

v = train speed, km/h

L_t = train length, m

$D = d/L_t$

(Naturally, if there is no locomotive, the above equation is simplified.) The total value of L_{Aeq24} may be estimated from the following:

$$L_{Aeq24} = 10 \log \left[(t_{ref}/T) \sum_{i=1}^{n} 10^{L_{axi}/10}\right] \qquad [2.7]$$

where

t_{ref} = 1 second

T is the total time period in seconds (= 86 400)

L_{axi} is the value of L_{ax} for each train

For a comprehensive method of estimating railway noise levels, see Myncke et al.[2.15]

The problem of railway noise in urban areas might be thought to be removed by placing the track underground. Whilst this reduces airborne sound propagation, it may cause another form of annoyance, as vibration induced in the ground by the passage of the trains may be transmitted into nearby building foundations. Older underground railways may cause a great deal of annoyance in neighbouring buildings, and this is very difficult to rectify (see Section 5.3). However, newer systems have been built using resiliently supported track and other means to reduce the vibration transmitted.

2.3.4 Shipping Noise

There are many types and sizes of ships and boats, ranging from large ocean-going vessels to small pleasure craft. Ocean-going ships for the most part operate far from land-based communities; however, their berthing and cargo-handling operations, with associated land-based transportation, together with the noise radiated from generators, fans, etc. whilst in port, can cause annoyance. The problem is exacerbated because of the need for 24-hour operations, the lack of shielding by buildings, etc. and the good transmission of sound energy over water.

The sound power levels of typical wharf cranes, gantries, fork-lift trucks and similar materials handling equipment frequently exceed

100 dB re 10^{-12}W. Another common noise source on wharves is associated with refrigerated containers holding perishable goods, which must be operated continuously—their sound power levels may be from 88 to 100 dB re 10^{-12}W.

In countries with major inland waterways used for commercial traffic, there are additional noise sources from the diesel engines of passing ships. There are standard methods of measuring ship noise emission; thus it is possible to set noise limits, at least for local vessels.[2.21, 2.22]

Pleasure craft may also cause considerable community noise annoyance; very powerful outboard mounted engines are used on some of these vessels, and high noise levels may be emitted unless controls are effected by regulatory authorities. Other sources are amplified music, used on 'party boats', and noise emitted during the launching and retrieval of trailer-mounted craft. Limitation of the times of operation of launching ramps is one method of reducing noise annoyance, although this may not be popular with amateur fishermen who wish to be on the water early in the morning. The best method of avoiding the problem is careful siting of launching ramps and associated car-parking areas and vehicle access routes.

2.3.5 Industrial Noise
Noise sources in industry are many and varied, and they are frequently the subject of control by environmental authorities. Many machinery operations involve the impact of metal parts on each other, e.g. gears, punch presses, etc. Bearings, conveyers, fans and associated gas flows, combustion processes, fluid flows, compressors, electric motors and generators, diesel and spark-ignition engines, all have characteristic noise emissions.

There are several ways in which industrial noise may be reduced at the source, which is the preferred method of control, since not only are nearby communities protected, but the employees' risk of permanent hearing damage is also reduced. In most cases, noise control is most effective if carried out as part of the initial machinery design; techniques such as cushioning impacts; using a smaller force over a longer time period; ensuring accuracy of gears; reducing the speed of rotating and moving parts in machines; reducing air and fluid flow velocities and general design of ducts, pipes, valves, etc. to smooth flow and to avoid turbulence; balancing rotating parts; using vibration damping materials and reducing the area of vibrating surfaces may all be used to reduce noise at the source.[2.23] In order to ensure that the best practical means of machinery noise control have been adopted, maximum noise levels should form part of the purchasing specification. Noise labelling, which gives the comparative noise levels of different models of the same type of machine under standardized test conditions, is being applied increasingly to assist purchasers in selecting the quietest plant and machinery.

Industrial noise impact on a community must be assessed on an individual basis. It is necessary to have information about the noise emission of all the plant and machinery to be used (preferably including one-third octave band power spectrum levels), and about the proposed building materials and construction, including the means of ventilation to be used. Calculations may then be made to determine the noise levels expected at appropriate receiver positions. (See Section 3.6.3). These levels may be compared either with Noise Limits set by the relevant authority, or with the existing background sound level in the area (See Section 3.6.2). For existing industry, the impact may be assessed by carrying out appropriate measurements, according to international or national standards.[2.24, 2.25]

2.3.6 People Noise

The activities of neighbours and their pets frequently cause annoyance, particularly in densely populated areas. Common sources are the use of high fidelity stereo systems at high sound levels, either with or without parties which may continue late into the night. Barking dogs may also annoy neighbours, particularly if they continue to bark over long periods of time. Burglar alarms, either on buildings or vehicles, cause considerable annoyance, unless they are fitted with time-controlled cut-off switches. Other neighbourhood noise sources are room air-conditioners, swimming pool filter pumps and automatic cleaners, lawn mowers, mulchers, home handyman power tools, etc. Many of these sources are not necessarily very loud, but it is the time of day (or night) when they occur that causes annoyance. They may all be the subject of legislative controls, but wherever possible, private resolution of the problem is preferable.

Where amicable solutions are not to be found then assessment of the extent of intrusion may be made by comparing the noise received by the neighbour when the alleged offending source is operating, with the background sound level when it is not (at the relevant time and place). Alternatively, maximum emission levels for each appliance may be set by the relevant authority.

2.4 SOUND PROPAGATION OUT OF DOORS

As discussed in Section 1.2 sound energy radiated from a point source operating in a free field propagates on a spherical wavefront with a continuously increasing radius. This is equivalent to a reduction of 6 decibels sound pressure level for each doubling of distance between source and receiver. If the source is linear, rather than a point, the radiation propagates on a cylindrical wavefront, and the attenuation rate is 3 decibels per doubling of distance. If the source is very large, and the receiving point is close to the source, the energy propagates on a plane wave front, and there is no attenuation of sound pressure level. The theoretical point-source and line-source sound pressure level reductions with distance are called *geometrical attenuation*. In practice, sound propagation may be affected by several other factors, and the additional attenuation which occurs is called *excess attenuation*. The major factors to be considered are absorption of sound by air, topography and barrier effects and meteorological effects.

2.4.1 Absorption of Sound by Air

That sound energy is absorbed by air was recognized by Rayleigh in the 19th Century,[2.26] and his theoretical explanation, in terms of viscosity and heat conductivity, was virtually unchallenged until the 1930s when Knudsen attempted to calibrate his new reverberation rooms at the University of California in Los Angeles.[2.27] Knudsen found that the relative humidity of the air in the chambers had a marked effect on their reverberation time (the time taken for sound to die away after the source had stopped: see Section 4.4.2.1). More recently, a comprehensive theory of sound absorption by air has been developed by Piercy and Embleton. Whereas thermal conductivity and viscosity, as postulated by Rayleigh, does absorb sound energy, this is only important at very high frequencies, above about 30 kHz (and thus unimportant in architectural acoustics and community noise studies, except when scale modelling techniques are used: see Section 6.8.1). Knudsen, Kneser and others postulated that the dependence of air absorption on the proportion of water vapour present could be interpreted in terms of the rate of adjustment of the internal equilibrium between vibrating and nonvibrating oxygen molecules, with the water molecules acting as a catalyst which greatly increased the reaction rate. If a periodic sound wave is propagated in air, a series of alternate adiabatic compressions and expansions occur; each compression is associated with a rise in temperature, and thus with an increase in the number of vibrating oxygen molecules; the expansion that follows allows the vibrational energy to be returned to the translating molecules, and if all of it is so returned, equilibrium is established. However, if the frequency of the sound wave is such that there is insufficient time for equilibrium to be re-established, the so-called *relaxation time*, an appreciable fraction of the vibrational energy is not returned as translational energy and this energy is extracted from the sound wave and dissipated as heat.[2.28] Although for pure oxygen, or pure dry air, this anomalous absorption takes place over a frequency band of about 3.8 octaves wide, centred below 100 Hz, the presence of small concentrations of water molecules shifts the absorption band to higher frequencies. In practice, at normal room temperatures and humidities, this absorption of sound due to relaxation of the oxygen molecules begins to be noticeable above 2 kHz. Piercy et al.[2.29] extended this theory in the 1970s to include relaxation of nitrogen molecules, which is also dependent on humidity. This is the main absorption mechanism for sound energy between about 200 Hz and 2 kHz, and amounts to about 2 to 3 dB per kilometre.[2.30] Several formulae have been published for the calculation of air absorption, including those of the American National Standards Institute, ANSI,[2.31] and the ISO.[2.32] A recent ISO draft proposal[2.33] has been shown to be more accurate than previous methods, although it does not agree exactly with experimental results.[2.34]

In practical cases, air absorption is negligible at low frequencies; at a distance of about 1000 m, it may cause a reduction of 2 to 10 dB at 1 kHz, depending on relative humidity, and it is increasingly important at high frequencies, which means that at distances of 1000 m (3 000 ft) or more there is little sound received above about 2 kHz.[2.35]

2.4.2 Ground Absorption

Most practical sound sources, apart from overflying aircraft, do not operate in a 'free field' away from reflecting surfaces. The source and receiving positions are usually located close to the ground, even in the absence of other surfaces and objects. This means that in addition to the direct sound wave, travelling through the air between source and receiver, there will be at least one reflected wave. The path of this reflected wave is calculated analogously to optical wave reflection. The *sound ray* (the normal to the wavefront at a particular point) is assumed to be reflected by the surface at an equal and opposite angle to that of the incident ray. It appears to originate at an *image source*, I, located

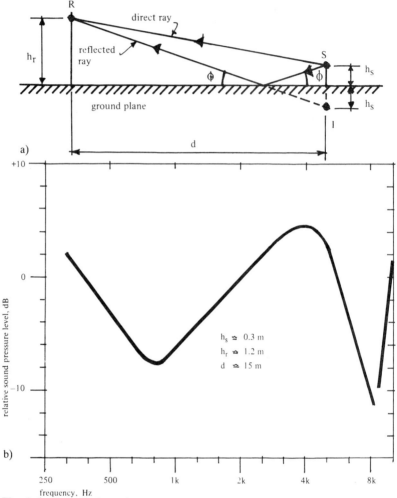

Fig. 2.5. Cancellation of sound energy at grazing incidence: a) geometry, b) typical theoretical spectrum received when propagated over grassland (after Embleton[2.30]).

equidistantly from the reflecting surface as the real source, S. The acoustical characteristic of the reflecting surface, its acoustic impedance, Z (See Section 1.2) will also affect the received sound energy. If the acoustic impedance of the surface exactly matched that of the air, all the energy impinging on the surface would be absorbed, and none would be reflected. The proportion of energy absorbed to that reflected depends chiefly on the ratio of the specific normal impedances for air and the ground surface, Z_1/Z_2, and the plane wave reflection coefficient, R_p, approximately:

$$R_p = (\sin \phi - \beta)/(\sin \phi + \beta) \qquad [2.8]$$

where ϕ = the grazing angle of incidence of the sound ray

$$\beta = Z_1/Z_2$$

For angles of ϕ small enough so that $\sin \phi$ is much less than β, R_p becomes effectively equal to -1. This implies a phase change on reflection and the reflected wave will then cancel the incident wave.[2.35] This simple geometrical acoustics approach is not really valid except for high frequency, short wavelength components. For low frequency sounds, parts of the wavefronts are incident on different parts of the surface with different incident angles, and Embleton describes this difference between simple theory and practice as the *ground wave effect*.[2.36] This ground wave plays a major role in low frequency sound propagation. The effect of cancellation of energy at grazing incidence is seen as an excess attenuation over a particular frequency band, often in the low frequencies (Fig. 2.5).

At large distances from a source there is a further component which becomes significant because it attenuates less rapidly than the direct and reflected waves. Piercy et al. describe this as a trapped surface wave which propagates through the air along a porous ground surface. The movement of air in and out of the surface pores is perpendicular to that of the normal sound wave in air. It is only excited for small values of ϕ but it appears to be the major carrier of environmental noise over long distances.[2.37] At large distances, of the order of 1 km (3 000 ft) when the source and receiver are close to the ground, the incident and reflected rays are almost at grazing incidence. This produces an almost complete cancellation between the direct and reflected waves and Embleton theories that the only sound received would consist of low-frequency ground and surface waves. However, since higher frequency sound energy is received at these distances in practice, it must be partly due to scattering by atmospheric turbulence.

The contribution of these various propagation mechanisms will vary according to the acoustical characteristics of the ground surface, and, in areas where the physical characteristics of the ground varies according to season, crops, snow cover, etc. variations in received sound level must be expected.

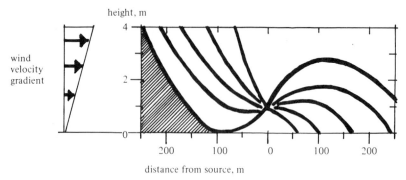

Fig. 2.6. Effect of wind gradient on sound propagation.

2.4.3 Meteorological Effects
The effect of relative humidity has already been discussed in relation to air absorption. Two other significant meteorological factors which affect sound propagation out of doors are wind and air temperature.

2.4.3.1 Wind effects
Sound waves move equally well with and against air movement. In an air-conditioning duct, for example, as much sound will be radiated from the air intake as from the air exhaust, in the absence of sound attenuating devices. In the open air, however, the effect of the differential wind velocity with height above ground must be taken into consideration. Wind near the ground has a lower velocity because of topographical irregularities, buildings, trees, and other obstacles. The further above the ground the less the wind velocity is influenced, and a wind gradient is formed as the velocity increases with height above the ground. If an omnidirectional source emits sound near the ground, the effect of the wind gradient is to bend some of the sound waves down towards the ground in the downwind direction, and away from the ground in the upwind direction. Thus more energy is received at a location downwind than in the absence of a wind gradient, and less upwind (Fig. 2.6).

2.4.3.2 Temperature effects
Under normal daytime conditions, the temperature of the air decreases with height above the ground—this is known as the *temperature lapse* condition. Since sound waves travel faster in warmer air than in cold

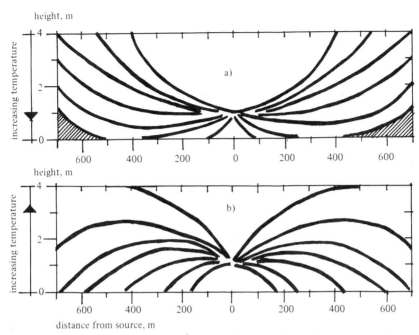

Fig. 2.7. Effect of thermal gradients on sound propagation, a) temperature lapse condition, b) temperature inversion.

air, during the day the sound waves from an omnidirectional source near the ground will tend to bend upwards, and less sound will be received at locations near the ground than if there were no temperature gradient. At night *temperature inversions* frequently occur in some climates; under these conditions, the air is colder nearer the ground than at greater heights above it. In this case the sound waves are bent downwards towards the ground, and higher sound levels are received (Fig. 2.7). Under extreme variations of temperature gradient, sound pressure level differences of up to 20 dB(A) have been measured.

Frequently both wind and temperature gradients exist simultaneously. However, it has been found that the influence of the wind gradient predominates over temperature effects when the velocity exceeds about 2 to 3 m/s (6 to 10 ft/s). Microclimatic changes, temperature gradients and turbulence tend to be prevalent near the ground and since many practical receiver and source locations are located in this area, quite large fluctuations in sound level must be expected if the propagation distance is over 100 m (300 ft) or even less. Elvhammer found that short term fluctuations over seconds or minutes are due to wind turbulence; variations over hours are due to diurnal temperature fluctuations, those occurring over days or weeks are due to weather changes, and level variations over months are due to ground and vegetation changes (particularly in climates where snow cover occurs).[2.38]

These changes in received sound levels due to meteorological variations cause problems when a project is being designed to meet noise control limits. There are also implications for compliance monitoring after completion. In some cases, the regulatory authorities specify particular meteorological conditions under which the noise control limits must be met (usually choosing those which will result in the highest levels at the receiving location); in other cases, a weighted average level, taking into account the typical variations in meteorological conditions is used.[2.39]

2.5 BARRIERS

It is common experience that a significant reduction of sound level occurs if the receiver is shielded from the source by topographical features, or by buildings. Man-made barriers have frequently been used in an attempt to attenuate noise, particularly from sources such as road traffic, aircraft engine testing, etc. However, the results have frequently been disappointing.

Geometrical considerations would lead to the assumption that sound energy falling on a barrier would be reflected away towards the source, and thus a receiver in the shadow of the barrier would not receive any sound. However, Huygens in the 17th Century assumed that every point on a wavefront acts as if it were a source and propagated a circular wave about itself. Since a wavefront is a locus of points in constant phases, all these circular waves, imagined emanating from each point in the given wavefront, will proceed outward in the same phase with one another. This is an explanation for the dispersal of the wavefront into the 'shadow' of an object. Fresnel and others developed the theory later, and in 1968 Maekawa published a simple, practical method of estimating barrier attenuation.[2.40]

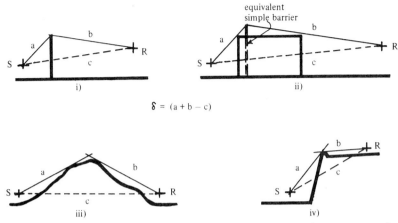

$$\delta = (a + b - c)$$

Fig. 2.8. Determination of Fresnel numbers for a given barrier geometry; i) simple barrier, ii) building on earth berm, iii) hill, iv) cutting.

It is necessary to determine the Fresnel number, N, of the barrier geometry (see Fig. 2.8), where

$$N = (2/\lambda)\delta \qquad [2.9]$$

λ = the wavelength of the sound being considered, m (ft)
δ = the path length difference between source and receiver, with and without the barrier [$\delta = (a + b) - c$], m (ft)

The attenuation for each band of sound, centred on the frequency for which λ was calculated may be found from Fig. 2.9. However, Maekawa's results apply to 'infinite' length barriers; if the barrier is not very long it is necessary to calculate the Fresnel numbers in the horizontal plane also, and to add up the contributions of all paths, logarithmically.

In practice, other factors, such as wind and scattering of sound into the shadow zone by turbulence, may reduce the effectiveness of a barrier. The partial cancellation of direct sound by out-of-phase ground reflected waves may also be prevented by the insertion of a barrier, thus reducing its insertion loss. Kurze points out that many barriers can not be represented by the thin screen assumed by theory—they may be wide earth berms, buildings, etc. and there may be multiple barriers formed by rows of buildings.[2.41] Daigle and others have compared theoretical attenuation of barriers in the presence of turbulence

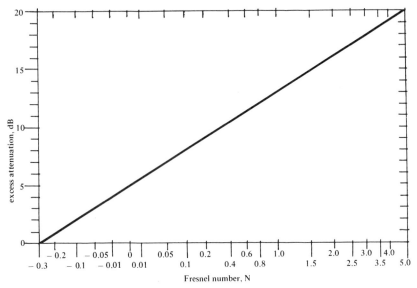

Fig. 2.9. Barrier attenuation vs Fresnel number (after Maekawa[2.40]).

and ground surfaces, with measurements. They found that for normal placement of barriers, close to either source or receiver, the prediction error may be of the same order as the insertion loss. Most simple methods of estimating barrier attenuation overestimate their effectiveness.[2.42, 2.43]

It is difficult to obtain more than about 10 dB(A) attenuation (for highway noise type spectra) from practical barriers in urban situations. This implies that it is not necessary to build massive barriers, since the sound transmission loss through the barrier need not exceed about 20 dB(A). (See Section 5.1). On the other hand, a building is itself usually a significant barrier, and the exposure of rear facades to road traffic or railway noise may be more than 20 dB(A) less than that of those facing the noise source directly.

Barriers have several disadvantages—alongside roads and railways they can be unsightly for both travellers and building occupiers, unless carefully designed, and they may also be a safety hazard along the side of highways. In addition, in warm climates, they may reduce cooling breezes. Some investigations have been made into the effect of barriers which are shaped to allow some air flow. However, it appears that their effectiveness as noise attenuators is severely compromised.[2.44]

It is frequently suggested that vegetation, such as a thick hedge or a belt of trees, may be used to attenuate noise, particularly that from road traffic. However, measurements do not confirm this, except in the high frequency range. Harris [2.45] found that a vegetative barrier could be expected to provide an insertion loss for highway noise of about 0.3 to 0.6 dB(A) per m (0.1 to 0.2 dB(A) per foot) and that a belt of mature vegetation of about 6 m, (20 ft) thickness could reduce the level from 4 to 6 dB(A). Although this is a barely perceptible reduction in noise level, some people have complained if relatively narrow belts of vegetation are removed—their reaction appears to be more related to the visual impact of the noise source. Aesthetic considerations should not be ignored, and if noise barriers are made to appear more attractive visually, either by their design, or by incorporating vegetative cover, they may have a greater noise annoyance reduction benefit than would be predicted from the actual acoustical attenuation provided.

2.6 OUTDOOR SOUND PROPAGATION PREDICTION

It will be realized from the foregoing, that the prediction of outdoor sound levels at some distance from a source should take into account many factors. The influence of some of these factors, such as wind and temperature gradients, varies over time. Authorities in some countries have prepared standard outdoor noise prediction methods to be used for legislative purposes, e.g. Scandinavia, Holland, West Germany. An ISO working group is preparing a model scheme.[2.46] It is designed for the prediction of the long-term average equivalent continuous sound level, $L_{Aeq, T,LT}$, and to give an indication of the average sound level under strong upwind or temperature lapse conditions. Five propagation problems are under study—air absorption, ground effects, including refraction and turbulence, screening including refraction and turbulence, other effects, such as fog, rain, snow, vegetation and source specifications, such as the effect of coherent sources, pure tones and propagation through built-up areas.

A revised draft scheme used in the Federal Republic of Germany, VDI 2714, Sound propagation outdoors,[2.47] includes the following factors:

$$L_S = L_W + DI + K_\omega - \Delta L_L - \Delta L_S - \Delta L_{BM} - \Delta L_D - \Delta L_G$$
$$- \Delta L_Z + \Delta L_R \qquad [2.10]$$

where

L_S = mean, down-wind received sound pressure level, dB(A)
L_W = mean sound power level, dB(A)
DI = directivity index
K_ω = correction for the reflection of sound energy
ΔL_L = correction for absorption of sound in air
ΔL_S = attenuation due to geometric spreading
ΔL_{BM} = correction for ground and downwind effects
ΔL_D = excess attenuation by absorption or scattering in vegetation
ΔL_G = excess attenuation by absorption or scattering in built-up areas
ΔL_Z = correction for screening effects
ΔL_R = correction for reflections from walls, etc. at receiving point.

2.7 SUMMARY

In this chapter community noise sources have been described, and methods of assessing the propagation of sound from transportation and industrial sources have been discussed. In the next chapter, specific planning methods that may be used to ensure the compatibility of different noise sources and land-uses are described.

REFERENCES

2.1. Schultz, T.J. Synthesis of social surveys on noise annoyance. *J.Acoust Soc.Amer.*64, 1978, pp 377-405.
2.2. Committee on the problem of noise, *Noise: final report* (Wilson Report) Cmnd 2056, Her Maj. Stat.Off. 1963.
2.3. Schultz, T.J. Some sources of error in community noise measurement. *Sound & Vib.* Feb. 1972, pp 18-27.
2.4.—*Acoustics—Description and measurement of environmental noise. Part 1— General Procedures.* AS 1055.1-1984, Standards Association of Australia.
2.5.—*Acoustics—description and measurement of environmental noise, Part 1— Basic quantities and procedures.* ISO 1996/1, International Standards Organisation.
2.6. Langdon, F.J. & Scholes, W.E. The traffic noise index—a method of controlling noise nuisance, *Building Research Station Current Paper*, CP 38/68.
2.7. Robinson, D.W. Towards a unified system of noise assessment. *J.Sound Vib.* 14, 1971, pp 279-298.
2.8. *Calculation of road traffic noise.* Dept. of Environment, UK. Her Maj. Stat.Off. 1975 (revised 1988).
2.9. Young, R.W. & Peterson, A. On estimating the noisiness of aircraft sounds. *J.Acoust.Soc.Amer.* 45, 1969, pp 834-838.

2.10. Hede, A.H. & Bullen, R.B. Aircraft noise in Australia: a survey of community reaction. *Nat.Acoust.Lab.Report. No.*88, 1982.

2.11. Gummlich, H.J. Aircraft noise abatement in the Federal Republic of Germany. *Internoise* 85, Munich, 1985 pp 1445-1448.

2.12. Vogel, A.O. *Noise zoning around airports in the Federal Republic of Germany according to the Air Traffic Noise Act.* Fed.Minister of the Interior, Feb.1981.

2.13. Gummlich, H.J. Noise abatement at general aviation airfields in the Federal Republic of Germany,*Internoise* 83 , Edinburgh, 1983, pp 639-642.

2.14. Fields, J.M. & Walker, J.G. Community response to railway noise in Great Britain, 10*th International Congress on Acoustics*, Sydney, 1980 C2-9.2.

2.15. Myncke, H., Cops, A. & Belder, P de. *Guide-line for the calculation of railway traffic noise*, Comm.of the Eur. Communities, 1980.

2.16. Kirschner, F. Koch, J.E. & Cohen, H.C. Lightweight vibration damping treatments for railroad vehicles. 11*th International Congress on Acoustics*, Paris, 1983, pp 129-132.

2.17. Nimura, T., Sone, T., Ebata, M. & Matsumoto, H. Noise problems with high speed railways in Japan. *Noise Con.Eng.* 5, 1975, pp 5-11.

2.18. Pfizenmaier, E. & King, W.F. On the relative importance of aerodynamic and wheel/rail noise sources generated by railway trains. *Internoise 84* , Hawaii, 1984, pp 177-180.

2.19. Cato, D.H. Prediction of enviromental noise from fast electric trains. *J. Sound Vib.* 46, 1976, pp 483- 500.

2.20. Roberts, C. The noise impact of freight train movements on residential communities in Western Australia. Proceedings, *Community Noise Conference*, Toowoomba, 1986, Aust. Acoust.Soc. pp 373-380.

2.21.—*Acoustics—Method for the measurement of airborne noise emitted by vessels on waterways, ports and harbours.* ISO 2922, International Standards Organisation.

2.22.—*Acoustics—Measurement of airborne noise emitted by vessels on waterways and in ports and harbours.* AS 1949-1988. Standards Association of Australia.

2.23. Berendt, R.D., Corliss, E.L.R. & Ojalvo, M.S. *Quieting: A practical guide to noise control*, NBS Handbook 119, US Dept. of Commerce, 1976.

2.24.—*Acoustics—Description and measurement of community noise environments, Part 1, Basic quantities and procedures, Part 2, Acquisition of data pertinent to land usage*, ISO 1996/1 & 2, International Standards Association, 1987.

2.25.—*Acoustics—Description and measurement of environmental noise, Part 1, General procedures, Part 2, Application to Specific Situations, Part 3, Acquisition of Data Pertinent to Land use*, AS 1055-1984 Parts 1, 2 & 3. Standards Association of Australia.

2.26. Rayleigh, J.W.S. *The Theory of Sound*, (reprinted) Dover, New York, 1945.

2.27. Knudsen, V.O. The effects of humidity upon the absorption of sound in a room, and a determination of the coefficients of absorption of sound in air. *J.Acoust.Soc.Amer.* 3, 1931, pp. 126-138.

2.28. Henderson, M.C. Sound in air: Absorption and dispersion, *Sound*, 2(6) Nov. 1963.

2.29. Piercy, J.E. Embleton, T.F.W. & Sutherland, L.C. Review of noise propagation in the atmosphere. *J.Acoust.Soc.Amer.* 61, 1977, pp. 1403-18.

2.30. Embleton, T.F.W. Sound propagation outdoors—improved prediction schemes for the 80's. *Noise Con.Eng.* 18,1982, pp 30-39.

2.31.—*Method for the calculation of the absorption of sound by the atmosphere.* ANSI S1.26-1978, Amercian National Standards Institute.

2.32.—*Acoustics—procedure for describing aircraft noise heard on the ground.* ISO 3891-1978. International Standards Organisation.

2.33.—*Acoustics—prediction of environmental noise- Part* 1: *calculation of the absorption of sound by the atmosphere.* First Draft Proposal ISO/DP 9613/1, 1986.

2.34. Davy, J.L. The absorption of sound by air, *Acoustics in the Eighties*, Proc. Aust.Acoust.Soc.Conf., Hobart, 1987, pp 47-54.

2.35. Embleton, T.F.W. Prediction of sound levels outdoors at distances up to 1 km. *Internoise 85* ,Munich. pp 449-452.

2.36. Embleton, T.F.W. Outdoor sound propagation. *Noise Con.* 87. Noise Control Foundation, New York, 1987, pp. 15-26.

2.37. Piercy, J.E., Embleton, T.F.W. & Daigle, G.A. Atmospheric propagation of sound: progress in understanding basic mechanisms. 11*th Int. Congress on Acoust.*, Paris, 1983, 8, pp 37-46.

2.38. Elvhammer, H. Checking external noise from industrial plants *Internoise* 77, Zurich, pp B 52-59.

2.39. *Acoustics—Description and Measurement of Environmental Noise. Part* 3: *Application to Noise Limits.* ISO 1996-3, International Standards Organisation, 1987.

2.40. Maekawa, Z. Noise reduction by screens. *App.Acoust.*1, 1968, pp 157-173.

2.41. Kurze, U.J. Noise reduction by barriers. *J.Acoust.Soc.Amer.*, 55, 1974, pp 504-518.

2.42. Daigle, G. Diffraction of sound by a noise barrier in the presence of atmospheric turbulence. *J.Acoust.Soc.Amer.*, 71,1982, pp 847-854.

2.43. Nicholas, J., Embleton, T.F.W. & Piercy, J.E. Precise model measurements versus theoretical prediction of barrier insertion loss in presence of the ground. *J.Acoust.Soc.Amer.*, 73, 1983, pp 44-54.

2.44. Maekawa, Z. & Osaka, S. Application of the line-integral method for designing a shaped noise barrier. *Internoise 86*, Cambridge, USA, 1986, pp 489-494.

2.45. Harris, R.A. Vegetative barriers: an alternative highway noise abatement measure. *Noise Con.Eng.* 27, 1986, pp 4-8.

2.46. Moerkerken, A. The state of the art in outdoor noise prediction schemes. *Internoise 86*, Cambridge, USA, 1986, pp 413-418.

2.47. Thomassen, H.G. Noise prediction, prognosis and planning. *Internoise 85*, Munich, 1985, pp 477-80.

CHAPTER 3

Community Noise and Planning

3.1 INTRODUCTION

Community noise problems could be avoided entirely if suitable land-use planning procedures were adopted universally; this type of noise problem arises solely through unsuitable juxtaposition of incompatible land-uses—for example, housing located near major highways, schools built near airports, etc. In an existing situation, if a community noise problem arises, it is difficult to find a solution using land-use planning procedures, and other methods, such as plant noise control, or building design and construction techniques must be used. Fortunately, there are many cases where it is possible to prevent future community noise problems arising by examining the noise impact of new developments on neighbouring areas, or by estimating the impact of noise from existing transportation routes and industrial enterprises on proposed neighbouring land-use developments. In this chapter, some guidelines will be presented to enable planners to assess the environmental noise impact of major roads, railways, airports and industrial developments on nearby sites. Different countries and states have varying legislative requirements and standardized methods for assessing the environmental impact of noise sources and naturally, where these exist, the planner must take cognisance of the exact requirements for their calculation. However, there are general principles, common to all methods, which will be discussed here, and these should assist planners and others to achieve practical resolution of their schemes.

3.2 NOISE CRITERIA FOR DIFFERENT DEVELOPMENTS

In general, the higher the activity noise levels in a development, the less sensitive it will be to noise from other sources. For example, a development such as a transport terminal, in which there are vehicles and materials handling equipment constantly in use, and where there is no need for unaided voice communication over medium to large distances, will not be incompatible with a major highway or an airport. On the other hand, a school, with need for good verbal communication indoors as well as its associated outdoor activity spaces, would have its operations severely affected in a similar location.

There are three principle criteria which may be used to assess the acceptability of ambient sound levels in different situations. Firstly, ambient sound levels should not interfere with desired audio communication; if this is to take place without electronic assistance, then the acceptable intruding sound levels are strictly related to the distance apart between speaker and listener, since human voice power is limited. Secondly, ambient sound levels should not cause annoyance to people carrying out their normal activities, such as when working, relaxing, studying, and particularly, when sleeping. The levels which are acceptable in these situations vary considerably from person to person, and, for any one person, from time to time. However, general guidelines are available for ambient sound levels which are usually acceptable to most people. The third criteria, the strictest of all, are those relating to theatres, auditoria and studios, where the perception of any sound other than that created in the performance itself is unwanted.

3.2.1 Ambient Sound Level Criteria for Unaided Speech Communication

In most cases, speech communication takes place inside buildings, and thus it is the ambient sound level inside the building that is of importance. This means that the acceptable ambient sound level outside (the community noise level) may be higher than that inside, because some attenuation of sound by the building structure may be assumed. However, if the building is naturally ventilated, the difference between outside and inside sound levels may be as little as 10 dB(A), for example, if 10% of a room's external surfaces consist of openings. (See Section 6.2.1). If normal domestic-type windows are closed, the attenuation may be of the order of 20 dB(A) or more. If the building is air-conditioned, and carefully constructed of substantial materials, much higher attenuation may be assumed. Thus, it is necessary for the land-use planner to make some assumptions regarding the building construction that exists or that is proposed for any particular development.

TABLE 3.1
Typical Recommended Maximum Ambient Sound Levels for Reliable
Communication at Different Distances Indoors

Communication distance, metres	Communication distance, feet	Ambient sound level dB(A)
4–9	13–30	35–40
2–4	6·5–13	40–45
2–4 (raised voice)	6·5–13	45–55

Some of the earliest work concerning acceptable ambient sound levels inside buildings was reported by Beranek, who was concerned primarily with determining design criteria for offices,[3.1, 3.2] and further studies were carried out over the next 15 years by Beranek and his colleagues.[3.3, 3.4] They found that mid-frequency background sounds are most important in determining the possibility of understanding conversational speech, and that this could be assessed objectively by determining the arithmetic average of the sound levels in the four one-octave bands centred on 500, 1000, 2000 and 4000 Hz, which is called the Preferred Speech Interference Level, PSIL. (An earlier proposal, the Speech Interference Level, SIL, omitted the 4000 Hz band.) Beranek also found that the overall sound level, expressed in phons, should not exceed the SIL by more than 22 units for acoustic conditions to be acceptable to office workers. He developed a series of curves, the Noise Criteria, or NC curves, which expressed this relationship; the Noise Criteria curve number is the same as the SIL number; e.g. NC 40 has an SIL of 40 and an overall level of 62 phons. Although the actual spectrum of the ambient sound level will affect the relationship between an NC rating and the overall sound level in dB(A), a common assumption is that the latter is 5 units higher.

Table 3.1 shows some recommended ambient sound levels for speech communication at different distances, derived partly from Beranek's earlier work, but expressed in terms of A-weighted decibels.

For planning purposes, if a building is to be naturally ventilated, and normal conversation is required at a distance of say 5 metres ($\cong$ 16 ft), the external ambient noise level should not exceed 45 to 50 dB(A). On the other hand, if it is accepted that telephone use will be slightly difficult, and that raised voices will be required for communication, the external noise levels may be 55 to 65 dB(A) and the building may still be naturally ventilated. If, as mentioned above, buildings are to be air-conditioned, and there are no associated exterior activities, higher external noise levels are acceptable. (For acceptable indoor sound level criteria for other building types see Chapter 6).

3.2.2 Ambient Sound Level Criteria for Dwellings

Sleeping areas are the most noise sensitive parts of dwellings. Research has shown that at certain levels, even though sleepers do not realise that they have been disturbed, their sleep patterns, as monitored using electroencephalograph records have been changed. As usual, there is considerable difference in the sensitivity shown by different subjects, but Griefahn recommends that an L_{Aeq} level of 40 dB(A) should not be exceeded indoors. This recommendation relates to fairly constant noise levels, with a range not exceeding about 10 dB(A), such as would be emitted by (attenuated) high-density road-traffic.[3.5] If possible, particularly if the range of levels is greater, as would be the case for low traffic flows, the L_{Aeq} level inside a bedroom should not exceed 25 dB(A), although in inner suburbs levels up to 35 dB(A) are acceptable.

In living rooms, typical criteria relate to radio and television programs. For speech, the dynamic range is about 30 dB(A) and for music, about 40 dB(A). If the maximum program levels are not to be too loud, it is necessary to limit the ambient sound levels so that the quietest part of the program can be heard. Recommended L_{Aeq} sound levels are 30 to 35 dB(A), with maxima of 40 dB(A) acceptable.[3.6]

For planning purposes, therefore, for naturally ventilated dwellings the external sound levels should not exceed 40 to 50 dB(A), daytime, and 35 to 45 dB(A) at night. (Ths is slightly more stringent than the maximum level of L_{dn} 55 recommended by the US EPA; this is equivalent to an L_{Aeq} of 55 during the daytime and of 45 dB(A) at night.)

3.2.3 Ambient Sound Criteria for Noise-sensitive Areas in Buildings

In auditoria, concert halls, studios, and the like, the intrusion of any noise other than that from the program is unacceptable, and it is the quietest parts of the program that provide the criteria for acceptable ambient sound levels. Usually, there will be an acoustical consultant involved in such buildings, and detailed ambient sound level criteria will be provided. However, as a general rule, the indoor sound levels should not exceed about 20 dB(A) for recording studios and 25 to 30 dB(A) for auditoria and concert halls at which audiences will be present. Most major buildings of this type will be air-conditioned, thus 30 dB(A) or more attenuation may be assumed between inside and outside. In addition to airborne sound criteria, these buildings are also sensitive to ground-borne vibration, and if the site is near existing or proposed underground railways, major highways, etc., this must also be taken into account.

3.3 GUIDELINES FOR LAND-USE NEAR ROADS AND HIGHWAYS

3.3.1 Measurement of Road Traffic Noise Levels

As discussed in Section 2.3.1, there are several well-known methods in use for the calculation or measurement of road traffic noise, but because of the effects of topography and meteorology (Sections 2.4.3, 2.5), if the road already exists and carries its designed traffic volume and mix, the most reliable method of assessing its impact on neighbouring land is to make detailed measurements over suitable time periods in relevant locations. However, this may be an expensive undertaking, particularly if a wide range of meteorological conditions is to be covered. Alternatively, sample measurements under known conditions may be made and corrections added to take into account expected variations.

3.3.2 Calculation of Road Traffic Noise Levels

If a new road is proposed, or an increased traffic flow is expected, calculations are necessary in order to assess the noise impact. Firstly, the estimated traffic flow rate and vehicle mix at a relevant time in the future, say over the next 10 or 15 years, should be determined. This information should be available from the highway authority. If the data is available only as an Average Annual Daily Traffic flow, AADT, it is necessary to estimate relevant hourly flow rates, usually relating to peak hours (or, if the $L_{10,\ 18\ \text{hour}}$ descriptor is used, the typical 18-hour flow rate). In urban areas there is often a bi-modal traffic flow distribution, with peak rates occurring in the morning and evening. However, as urban areas increase in size, the times of low traffic flow rates tend to reduce, and there is less fluctuation over the 24 hour period. Each case should be considered on its merits. Fortunately the emitted noise levels vary only with the logarithm of the traffic flow rate, thus small inaccuracies in traffic growth rate assumptions are relatively unimportant. For example, if the traffic flow rate is doubled, from say 1000 to 2000 vehicles per hour, the increase in emitted noise level is only about

3 dB(A), L_{Aeq}. (However, the same incremental difference occurs for a doubling from 100 to 200 vehicles per hour; thus the lower the initial flow rate, the more critical is an accurate estimation of traffic volume). Information concerning the expected percentage of 'heavy' vehicles is also required. A change in emitted noise levels of 6 to 12 dB(A) (depending on the calculation method used) is expected as the percentage of heavy vehicles increases from 0 to 40%. For some detailed prediction methods, information concerning the expected vehicle speed, the road surfacing material and the road gradient is also required.

The distance of the relevant receiving location(s) from the road, and any topographical or other barriers that exist or that will be erected are also required for the calculation of noise levels. Again, since the emitted noise levels attenuate according to the logarithm of the distance, small errors in distance estimation are not critical for planning purposes.

One method of estimating traffic noise levels was developed in the United Kingdom, where it is used for assessing eligibility for monetary compensation to enable building modifications to be carried out to reduce traffic noise intrusion.[3.7] An example of the calculation procedure using the UK method follows:

Example

A proposed new road will carry a maximum of 2 000 v/h, with a maximum of 15% heavy vehicles. The road gradient is level and the expected average vehicle speed is 70 km/h (44 mph). It is required to estimate the maximum noise level, hourly L_{10}, at a position 30 m (98 ft) from the edge of the nearside kerb, and 3·5 m (11.5 ft) above the ground.

1. Basic noise level hourly.

$$L_{10} = 41·2 + 10 \log q$$

(where L_{10} = the noise level exceeded for 10% of the time period, dB(A) ·
q = the hourly vehicle flow, vehicles/h)

$$L_{10} = 41·2 + 10 \log 2{,}000$$
$$= 74·2 \text{ dB(A)}.$$

2. Correction for heavy vehicles and vehicle speed.

$$\text{Correction} = 33 \log [v + 40 + (500/v)] + 10 \log [1 + (5p/v)] - 68·8$$

(where v = vehicle speed, km/h
p = percentage heavy vehicles)

$$= 33 \log [70 + 40 + (500/70)] + 10 \log [1 + (5 \times 15/70)]$$
$$- 68{\cdot}8$$
$$= 68{\cdot}27 + 3.16 - 68{\cdot}8$$
$$= 2.63 \text{ dB(A)}$$

3. Correction for gradient.

Correction $= 0{\cdot}3\text{G dB(A)}$
(where G = gradient, %)
$$= 0{\cdot}3 \times 0$$
$$= 0 \text{ dB(A)}$$

4. Correction for distance (assuming hard ground)

Correction $= -10 \log (d'/13.5)$
(where $d'=$ shortest slant distance to the receiving point from the effective source position, and the effective source is assumed to be 3.5 m (11.5 ft) from the nearside kerb and 0.5 m (1.6 ft) high, typical of low-level vehicle exhausts. Since the receiving point is 3.5 m (11.5 ft) above the ground, it is 3 m (9.8 ft) above the effective source height and 33.5 m (109.9 ft) away horizontally).

$d' = (33{\cdot}5^2 + 3^2)^{1/2}$
$ = 33{\cdot}63 \text{ m (or 110.34 ft)}$

Correction $= -10 \log (33{\cdot}63/13{\cdot}5)$ [or $-10 \log (110{\cdot}34/44{\cdot}3)$]
$ = -3{\cdot}96 \text{ dB(A)}$

5. Total estimated sound level at receiving position:

$(1 + 2 + 3 + 4)$
$L_{10} = 74{\cdot}2 + 2{\cdot}63 + 0 - 3{\cdot}96$
$\phantom{L_{10}} = 72{\cdot}9 \text{ dB(A), say 73 dB(A)}$

This L_{10} level of 73 dB(A) is approximately equivalent to an L_{Aeq} level of 70 dB(A). If this land was to be used for a naturally ventilated building the internal noise levels would be approximately 60 dB(A), which would be unacceptable for dwellings, offices, etc. However, it would be satisfactory for many industrial type uses.

For preliminary planning purposes, a simplified equation may be used to estimate L_{A10} or L_{Aeq}. The noise level emitted by a vehicle at speeds less than about 80 km/h (50 mph) is primarily related to its engine speed.

If a vehicle is constantly accelerating and decelerating in urban traffic, its engine speed is not simply related to its road speed; in this situation, the term for vehicle speed has been found not to be relevant. Burgess developed some empirical prediction equations, appropriate for built-up areas where traffic flow is interrupted, as follows:[3.8]

$$L_{Aeq} = 55 \cdot 5 + 10 \cdot 2 \log Q + 0 \cdot 3 \, p - 19 \cdot 3 \log d \qquad [3.1]$$
$$L_{A10} = 56 + 10 \cdot 7 \log Q + 0 \cdot 3 \, p - 18 \cdot 5 \log d \qquad [3.2]$$

where

Q = the traffic flow rate, vehicles per hour
p = percentage ' heavy' vehicles
d = distance from the centre of the nearside flow, m.

These simplified equations may be used, for example, to estimate the width of the buffer zone required between a road and a residential area, or the effect of an increase or decrease in traffic flow resulting from a traffic management scheme.

Example
What is the width of the buffer zone required between an urban road carrying an average of 1000 vehicles per hour, daytime, with 10% heavy vehicles, if the traffic noise level inside living rooms facing the street is not to exceed 40 dB(A), L_{Aeq}.

Assume the attenuation provided by the facade, with open windows, will be 10 dB(A); therefore, the maximum allowable external level will be 50 dB(A). Using the Equation 3.1 for L_{Aeq}:

$$L_{Aeq} = 55 \cdot 5 + 10 \cdot 2 \log Q + 0 \cdot 3 \, p - 19 \cdot 3 \log d$$

Therefore,

$$19 \cdot 3 \log d = 55 \cdot 5 + 10 \cdot 2 \log Q + 0 \cdot 3 \, p - L_{Aeq}$$
$$\log d = [(55 \cdot 5 + 10 \cdot 2 \log 1000 + 0 \cdot 3 \times 10) - 50]/19 \cdot 3$$
$$d = 106 \text{ metres } (\cong 348 \text{ ft})$$

Thus the required buffer zone, assuming line-of-sight propagation, will need to be slightly over 100 m, (330 ft) wide. Alternatively, if the land is reasonably level, and the buildings are single storied, it may be more economical to build a roadside barrier and thus enable the width of the buffer zone to be reduced

---

3.3.3 Guidelines for Estimating Barrier Attenuation

The attenuation of sound provided by natural or man-made barriers may be estimated. (This is the excess attenuation, which is added to the geometrical spreading attenuation of 3 or 6 dB per doubling of distance between source and receiver, for a line or point source respectively.) As discussed in Section 2.5 in practice there are many factors which reduce the theoretical effectiveness of barriers. For critical applications, careful analysis of the expected barrier attenuation is necessary.[3.7, 3.9, 3.10] However, for preliminary planning purposes, the following approximate relationships may be applied:

$$N_f = 10 \log 20 \, X_f \qquad\qquad [3.3]$$

where

N_f = the excess attenuation of the barrier, in dB, for a band of sound centred on frequency f

$$X = \frac{2\left[R\left(\sqrt{1 + \left(\dfrac{H}{R}\right)^2} - 1\right) + D\left(\sqrt{1 + \left(\dfrac{H}{D}\right)^2} - 1\right)\right]}{\lambda\left(1 + \left(\dfrac{H}{R}\right)^2\right)} \qquad [3.4]$$

where

R = distance between source and barrier, m (ft)
D = distance between barrier and receiver, m (ft)
H = effective height of the barrier, m (ft)
λ = wavelength of the centre frequency of the band of sound, m(ft)

In cases where D $\gg$ R $\geqslant$ H, the above relationship may be simplified to:

$$X_f = H^2/(\lambda R) \qquad\qquad [3.5]$$

Fig. 3.1 illustrates the geometry involved. It should be noticed that the effective height of the barrier, H, is not usually as great as the height of the barrier above the ground. If the barrier is not a thin screen, but consists of an earth mound, or a hill, etc., an approximate value for H must be assumed. Figure 2.8 shows a suggested approach, although care must be taken, particularly if the source and/or receiver is located close to the barrier.

The above equations predict an increased attenuation of 3 dB for each halving of wavelength (doubling of frequency). However, as discussed

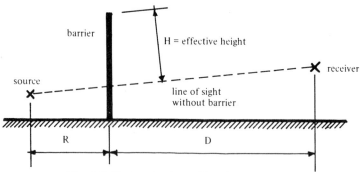

Fig. 3.1. Determination of barrier geometry.

earlier, because of scattering due to turbulence, wind and temperature gradients and the loss of ground absorption, the theoretical results tend to overestimate the actual attenuation that will be obtained. If the barrier is of limited horizontal extent, the sound paths around the ends must also be calculated, and the total sound level determined by logarithmic addition. It is difficult to achieve more than about 10 dB(A) attenuation of highway noise for practical height barriers. When either or both the source and receiver are elevated, barriers are of little use.

Using the same example as before, an estimate may be made of the reduction in required buffer zone between the road and the dwellings.

Example
For the same conditions as the previous example, calculate the required width of the buffer zone between the road and the dwellings, if a barrier with an estimated attenuation of 10 dB(A) is erected along the length of the road.
The calculation may be made as follows:

$$19 \cdot 3 \log d = 55 \cdot 5 + 10 \cdot 2 \log Q + 0 \cdot 3\, p - 50 - 10$$
$$\log d = [(55 \cdot 5 + 10 \cdot 2 \log 1000 + 0.3 \times 10) - 60)]/19 \cdot 3$$
$$d = 32 \cdot 2 \text{ metres } (\cong 106 \text{ ft})$$

This is a saving of nearly 70 m ($\cong$ 230 ft) width of buffer zone. In order to decide whether this is the most economical solution, the saving in land costs must be compared with the cost of construction of the required barrier. The approximate height of the required barrier may be found from the simplified Equations 3.3 and 3.5.

Assume that the barrier will be built at a distance of 6 m ($\cong$ 20 ft) from the traffic stream and that the source height is 0.5 m (1.6 ft) above the road. If the nearest dwelling is 32 m (105 ft) from the traffic, it will be approximately 26 m (85 ft) from the barrier. Assume the receiver height is 3 m ($\cong$10 ft) above the ground, and that the performance of the barrier at 1000 Hz will give an indication of its overall attenuation in dB(A), taking into account a typical A-weighted traffic noise spectrum. The required attenuation, N, is 10 dB.

$$\text{Since N} = 10 \log 20X$$
$$10 = 10 \log 20X$$
$$\log 20X = 10/10 = 1{\cdot}0$$
$$20X = 10, \text{ therefore X} = 0{\cdot}5$$

Thus since $\quad$ $X = H^2/\lambda R$
$$0{\cdot}5 = H^2/(0{\cdot}344 \times 6)$$
$$H = (0{\cdot}5 \times 0{\cdot}344 \times 6)^{1/2} \text{ [or H} = (0.5 \times 1{\cdot}13 \times 20)^{1/2}$$
$$H = 1{\cdot}02 \text{ metres} \qquad \cong 3{\cdot}4 \text{ ft]}$$

H is the effective height of the required barrier; geometry must be used to calculate the actual height above the road, which will be 1.5 m ($\cong$ 5 ft)

In the above example, the source height was taken as only 0.5 m (1.6 ft), which is representative of the heights of automobile and other low level exhausts, and of typical engines above the ground. However, if there is a significant proportion of commercial vehicles with vertical exhausts, a higher source position should be used in the calculations. Also, it was assumed that the barrier would be built reasonably close to the centre of the nearside traffic stream. If the road is divided by a wide median strip, it would be appropriate to calculate the effectiveness of the barrier with respect to the far lanes of traffic also.

3.4 GUIDELINES FOR ASSESSING COMPATIBLE LAND-USE NEAR AIRPORTS

3.4.1 Measurement of Aircraft Noise

It is extremely difficult to determine the aircraft noise exposure of a site simply by making measurements of overflying aircraft. This is because of the discrete nature of aircraft noise events and their variability according to aircraft type, flight path, loading and general operating conditions. Even if there is only one runway and a preferred operational flight path, it is not unusual to observe considerable differences in the actual route flown by different pilots. If the actual aircraft noise levels for a particular site are required it is necessary to install some type of permanent or long-term monitoring system. Many such systems are installed near major airports, and they are often linked to the airport control tower computer system so that individual aircraft noise emissions may be identified.

3.4.2 Calculation of Aircraft Noise Exposure

This is normally done in two stages, as discussed in Section 2.3.2. The first stage is to determine whether or not aircraft noise intrusion is sufficiently frequent to affect a given site or building; the second stage is to determine the expected maximum aircraft flyover noise levels. In some countries it is possible to obtain an aircraft noise exposure map from the aviation control authority. This will normally show contours of approximately equal exposure and will be accompanied by guidelines for siting different building types within particular noise exposure zones. For some buildings even occasional aircraft noise intrusion is not acceptable, for example a major concert hall, in which case consultation with the relevant aviation authority to determine the actual maximum flyover levels expected at the site is recommended. However, for most buildings, occasional aircraft noise intrusion will not be of great consequence, which is fortunate, because such occurrences are possible, and even probable, well outside the areas nominally affected by a given airport.

The building site should be located on the aircraft noise exposure map to determine whether the proposed use is compatible. If this is found to be the case, the required aircraft noise attenuation to be provided by the building can be determined, according to the distance of the site from the airport runways.

Example
(This example is related to the method used in Australia, based on AS 2021-1985 *Acoustics—Aircraft noise intrusion—building siting and construction*.[3.11] In other countries different assessment procedures may be used.)

It is proposed to build a new school within 10 km (6.25 miles) of a

TABLE 3.2
Aircraft Maximum Flyover Noise Levels for Design
Purposes

Aircraft type	Take off 3·5 km distance, 0·5 km sideline	Landing 2 km distance, 0·5 km sideline
B 747	84	80
DC 10	84	76

major international airport. An aircraft noise exposure map showing Australian Noise Exposure Forecast zones around the particular airport is available, and the site is found to be within the 20 to 25 ANEF zone. (See Fig. 3.2) This is 'conditionally acceptable' for a school. The next stage is to determine the 'indoor design sound levels for aircraft noise reduction assessment'; these are given as 50 dB(A) for libraries and study areas, 55 dB(A) for teaching and assembly areas and 75 dB(A) for gymnasia and workshops. The most critical level is 50 dB(A). (These levels are much higher than those considered acceptable for continuous noise, for example such as that propagated from a nearby road; this is because aircraft noise events are discrete. The ANEF zoning screen ensures that the number of such events should not be excessive for a given building use.)

As the airport is a major international one, large, wide-bodied jet aircraft, such as Boeing 747 and McDonnell-Douglas DC 10 will be using it. If the site is 2 km (1.25 miles) from the nearest touch-down point, 3.5 km (2.2 miles) from the nearest start-of-roll for take-off point and 0.5 km (0.73 miles) to the side of the extended runway centreline, the maximum aircraft noise levels expected, in dB(A), are as shown in Table 3.2.

For this site, the take-off noise levels are the highest, at 84 dB(A). The required aircraft noise attenuation will be (84−50) = 34 dB(A) for the most sensitive parts of the school and (84−55) = 29 dB(A) for the general teaching areas. This attenuation requirement is compatible with standard forms of building construction, unless the rooms are to be naturally ventilated, in which case the overall envelope attenuation with open windows will be only about 10 dB(A). Thus, in order to comply with the guidelines of the Standard, it will be necessary to use mechanical ventilation or air-conditioning for a school at this site. A further section of the Standard includes a method for determining appropriate forms of construction for roof/ceiling, walls and windows and doors (See Section 6.2.1).

3.5 GUIDELINES FOR ASSESSING COMPATIBLE LAND-USE NEAR RAILWAYS

3.5.1 Measurement of Railway Noise
Because of the wide variety of rail-bound vehicles in use in different countries, it is difficult to generalise on expected noise emission from railways. Measurement of the noise of individual train pass-bys, covering the range of locomotives and rolling stock, and the range of speeds expected, should be made to develop a data base for a given system. This data may then be used to predict a general noise descriptor, such as $L_{Aeq,24}$, for sites at different distances from the track. The usual estimates of additional attenuation provided by barriers and topography should also be made.

3.5.2 Calculation of Railway Noise Impact
As mentioned above, the impact of railway noise on a nearby site may be estimated if information is available regarding the maximum passby levels of the trains using the tracks, and their frequency of operation. In the example given below it is assumed that the impact is required in terms of the equivalent energy level over a typical 24-hour period, $L_{Aeq,24}$. If the site is to be used for a limited time period within 24 hours then the data relevant for this time period should be used. Refer to Section 2.3.3. for a discussion of the derivation of this method.

Example
Assume that the maximum passby noise levels, measured 20 metres ($\cong$ 66 ft) from the track, from the two types of train using a particular railway are 90 and 87 dB(A) respectively. It is expected that over a 24 hour period the trains emitting the higher level will have an average frequency of 5 times per hour, and the others will have a frequency of 15 times per hour. The trains are 300 m ($\cong$ 980 ft) long and travel at an

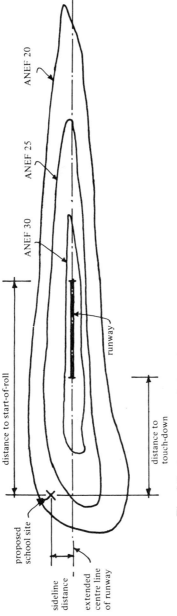

Fig. 3.2. Assessment of aircraft noise intrusion, location of site *vis à vis* airport.

average speed of 60 km/h (37.5 mph). What will be the L_{Aeq24} level at a site located 100 m ($\cong$ 330 ft) from the track?

$$L_{AxA} = L_{AmaxA} + 10 \log \frac{L_t}{v} - 10 \log \left[\frac{4D}{4D^2 + 1} + 2 \tan^{-1}\left(\frac{1}{2D}\right) \right] + 10 \cdot 5$$

where

$$D = \frac{d}{L_t} = \frac{(100 - 20)}{300} = 0 \cdot 27 \qquad \left(\text{or } D = \frac{(330 - 66)}{980} \cong 0 \cdot 27 \right)$$

$$L_{AxA} = 90 + 10 \log \frac{300}{60}$$

$$- 10 \log \left[\frac{4 \times 0 \cdot 27}{4 \times 0 \cdot 27^2 + 1} + 2 \tan^{-1}\left(\frac{1}{2 \times 0 \cdot 27}\right) \right] + 10 \cdot 5$$

$$= 90 + 6 \cdot 99 - 20 \cdot 95 + 10 \cdot 5$$

$$= 86 \cdot 5$$

$$\cong 87 \text{ dB(A)}$$

$$L_{AxB} = (L_{AxA} - 3) = 84 \text{ dB(A)}$$

(Note the train lengths and speeds were assumed to be the same)

Train A has a frequency of 5 times per hour, thus in a 24 hour period there will be 120 passbys; Train B has a frequency of 15 times per hour, and there will be 360 passbys in a 24 hour period. Therefore:

$$L_{Aeq24} = 10 \log (t_{ref}/T) \sum_{i=1}^{n} 10^{L_{axi}/10}$$

$$= 10 \log [1/86,400 \times (120 \times 10^{87/10} + 360 \times 10^{84/10})]$$

$$= 10 \log [1 \cdot 1574 \times 10^{-05} \times (6 \cdot 01 \times 10^{10} + 9 \cdot 04 \times 10^{10})]$$

$$= 62 \cdot 4, \text{ say } 62 \text{ dB(A)}.$$

A level of 62 L_{Aeq24} would be unacceptable for naturally ventilated dwellings, but it would be quite suitable for air-conditioned buildings such as offices, provided that the construction chosen had the necessary sound attenuating properties.

3.6 GUIDELINES FOR ASSESSING ACCEPTABLE LAND-USE NEAR INDUSTRY

3.6.1 Measurement of Industrial Noise
If the industrial development already exists, measurements of its noise emission may be made at suitable locations. However, care must be

taken that the measurements are made under the relevant meteorological conditions, particularly if the receptor locations are more than 30 to 100 m (100 to 330 ft) from the source. (See Section 2.4.3) If the process operations vary during a 24 hour period—for example, if the night shift does not include all the daytime operations, or if there is no night-time operation at all—this should be taken into account when deciding on the relevant time periods for carrying out the noise measurements.

The results of the measurements, using the relevant descriptor (which may be in terms of $L_{Aeq, t}$, where t may be one or more different periods e.g. daytime and night-time), may then be assessed in terms of land-use compatibility for different purposes (See Chapter 6). As discussed before, the method of ventilation of any proposed building will have a critical effect on the acceptability or otherwise of a particular noise level. If natural ventilation is to be used the attenuation of the building envelope cannot be assumed to be higher than about 10 dB(A). Alternatively, the acceptability of the industrial noise emission may be assessed according to Noise Limits set by the relevant authority. If the Noise Limits are exceeded then noise control measures must be instituted.

In some circumstances, the relevant authority may require noise monitoring to be carried out. Where there are many enterprises, each of which is considered as a noise source, it may be necessary to monitor the noise emission of each enterprise within its own boundaries. The permissible noise levels are then determined by calculating the attenuation of sound between the source and noise-sensitive receiving locations, taking into account geometrical spreading, man-made and natural (topographical) barriers, and meteorological effects.

3.6.2 Measurement of Background Noise Levels

In some instances, the acceptability of an industrial development is assessed with respect to the pre-existing background noise levels at receiving locations, particularly for residential land-uses. If the industrial processes are continuous over the 24 hour period, the most critical assessment period for residential purposes is usually at night. This is because general community noise levels are often lower at night, and, in addition, the effects of temperature inversion may enhance the received noise levels at locations more than 30 to 100 m (100 to 330 ft) from the source. (See Section 2.4.3).

Wherever possible, measurements of the background sound level should be made with the source under investigation (in this case the industrial development) not operating. If measurements are repeated at the same location with the source operating, an assessment may be made of its impact, if any, on the receiving area. If the background sound level without the source operating and the industrial noise level are described in terms of L_{Aeq} values, differences of 5 dB(A) or less are usually considered to be of marginal significance; however, if the industrial noise exceeds the background noise level from all other sources at a site by more than 5 dB(A), community noise annoyance and complaints may be expected to occur. Alternatively, the background noise may be described in terms of L_{90} or L_{95} and the industrial noise in terms of L_{10}. Again, differences of 5 dB(A) or less are considered to be of marginal significance.

Characteristics of an industrial noise which make it more noticeable, such as tonal components or impulsiveness, tend to make it more annoying. In some assessment schemes 'correction factors' are used to make allowance for this, e.g. 5 dB(A) may be added to the measured level if a prominent tonal component is audible.[3.12] If the sound has an impulsive character, additional measurements using an 'Impulse' time-weighting should be made, and if these levels are higher than those using the 'Fast' time-weighting, they should be used for the assessment of acceptability.

3.6.3 Prediction of Industrial Noise Levels

If it is required to assess the noise impact of a proposed industrial development it is necessary to determine the expected noise emission from the various machines and plant that will be included. If possible, the data should include the one-third-octave band sound power levels of all the sources, and their times of operation. From this information the total sound level inside the industrial building(s) can be estimated. Additional sources external to any building(s) must also be determined.

The next step is to estimate the sound attenuation provided by the building envelope. For this it is necessary to have information concerning the materials to be used for the walls and roof/ceiling, and whether or not natural ventilation is to be used. In many cases industrial buildings have large, permanent access openings and these allow sound to be transmitted outside with little attenuation. (See Section 6.7).

The attenuation of sound due to distance (geometrical spreading) must then be estimated. If the receiving location is far from the source the latter can be considered to radiate energy as a point source, and a maximum attenuation rate of 6 dB per doubling of distance can be assumed. However, if the source is large compared to distance between

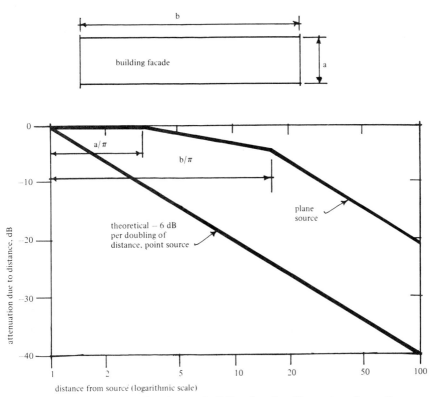

Fig. 3.3. Sound propagation from a building facade—illustration shows theoretical attenuation for a = 10 and b = 50 (metres or feet).

it and the receiver, little or no attenuation will occur due to geometric spreading. A rule-of-thumb estimate is that no attenuation will occur (plane wave propagation) within a distance of a/π from the facade of a building, where a is the smaller dimension of the facade, and that beyond this distance the building will act as a line source, with an attenuation rate of 3 dB per doubling of distance up to a total distance from the facade of b/π where b is the larger dimension. The point-source attenuation rate then takes place beyond this distance. (See Fig. 3.3).

It should be mentioned that this is indeed a simplified method of assessing attenuation of sound with distance from the source. For more information regarding some of these methods see the references.[3.13, 3.14, 3.15] Some industries essentially operate in open-air, for example oil refineries, etc: in this case it is necessary to calculate the expected total sound power level from all the sources and to assume an 'acoustic centre' from which they propagate.[3.16, 3.17]

Example
The impact of a proposed new industrial building on a residential area located 50 m (164 ft) away is to be assessed. The operation will be over 24 hours, 7 days per week. The total sound pressure levels at 1 m (3.3 ft) from the facade have been estimated to be 75 dB(A); the sound will be broad-band and will not have prominent tonal components or impulsive characteristics. The facade nearest the residential area has the dimensions of 50 m x 10 m (164 ft x 33 ft) high.

1. Attenuation rate from facade to $a/\pi = 0$ dB. Therefore from facade to $10/\pi = 3.2$ m (10.5 ft) attenuation = 0 dB
2. From $10/\pi$, 3.2 m (10.5 ft) to $50/\pi$, 15.9 m (52 ft) attenuation rate will be 3 dB per doubling distance (dd).
3. From $50/\pi$, 15.9 m (52 ft) attenuation rate will be 6 dB per dd.

Therefore the total attenuation, at 50 m (164 ft) will be:

$(1 + 2 + 3) = 0 + 6·9 + 9·9 = 16·8$ dB(A)

and the sound level at residential area will be

$(75 - 16·8) = 58·2$, say 58 dB(A).

This would be unacceptable, particularly for a 24-hour operation. If a maximum acceptable sound level is assumed to be 35 dB(A), the level at 1 m (3.3 ft) from the facade should not exceed $(35 + 16.8) = 51.8$, say 51 dB(A). This will require noise control measures providing an attenuation of 24 dB(A) to be carried out either within the industrial building, or by changes to the building's construction.

The above calculations have not taken into account the spectrum of the sound, ground and air absorption, or adverse conditions which may occur during downwind sound propagation, temperature inversions, etc. However, they may be used to make a first estimate of land-use compatibility for planning purposes.

3.7 SUMMARY

In this chapter some examples have been given of methods of estimating the compatibility of different land-uses with respect to noise emitted by various transportation and industrial sources. They are not intended to be exhaustive; rather they are methods that may be used by planners and developers to make a first estimation of possible acoustical problems that may arise in new or existing developments.

REFERENCES

3.1. Beranek, L.L. Criteria for office quieting based on questionnaire rating studies. *J.Acoust.Soc. Amer.* 28, 1956, pp.833-852.

3.2. Beranek, L.L. Revised criteria for noise in buildings. *Noise Control* 3, 1957, pp 19-27.

3.3. Cavanaugh, W.J., Farrell, W.R., Hirtle, P.W. & Watters, B.G. Speech privacy in buildings *J.Acoust.Soc.Amer.* 34, 1962, pp 475-492.

3.4. Beranek, L.L., Blazier, W.E. & Figwer, J.J. Preferred Noise Criteria Curves and their application to rooms. *J.Acoust.Soc.Amer.* 50, 1971, pp 1223-1231.

3.5. Griefahn, B. Research on noise-disturbed sleep since 1973. *Noise as a Public Health Problem*, Freiburg, Amer.Speech & Hearing Assoc. Reports, 10, 1980, pp 377-390.

3.6.—*Acoustics—Recommended design sound levels and reverberation times for building interiors. AS* 2107-1987, Standards Association of Australia.

3.7.—*Calculation of Road Traffic Noise*, Dept. of the Environment, Welsh Office.1975. Her Maj.Stat.Office (revised 1988).

3.8. Burgess, M.A. Noise prediction for urban traffic conditions—related to measurements in the Sydney Metropolitan Area. *App.Acoust.* 10, 1977, pp 1-7.

3.9. Maekawa, Z. Noise reduction by screens. *App.Acoust.*1, 1968, pp 157-173.

3.10. Kurze, U.J. Noise reduction by barriers, *J.Acoust Soc. Amer.* 55, 1974, pp 504-518.

3.11.—*Acoustics—Aircraft noise intrusion—building siting and construction.* AS 2021-1985. Standards Association of Australia.

3.12.—*Acoustics—Description and measurement of environmental noise, Part 1: General procedures.* AS 1055- 1984, Part 1, Standards Association of Australia.

3.13. Thomassen, H.G. Noise prediction, prognosis and planning. *Internoise 85*, Munich, 1985, pp 477-480.

3.14. Driscoll, D.A. NOISECALC: A computer program for sound propagation calculations. *Noise Con.Eng.* 25, 1985, pp 88-92.

3.15. Moerkerken, A. The state of the art in outdoor noise prediction schemes. *Internoise 86*, Cambridge, USA, 1986, pp 413-418.

3.16. Marsh, K.J. The CONCAWE model for calculating the propagation of noise from open-air industrial plants. *App.Acoust.* 15, 1982, pp 411-428.

3.17.—*Acoustics—Determination of sound power levels of multi-source industrial plants for the evaluation of the sound pressure levels in the environment— engineering method.* ISO/DIS 8297-1988, International Standards Organisation.

CHAPTER 4

Room Acoustics

4.1 INTRODUCTION

There are two main factors to be considered in the acoustical design of buildings: firstly, wanted sound, be it speech or music, must be heard under the best possible conditions, and secondly, unwanted sound, or noise (which may include speech or music) must be prevented from entering or leaving the building. There is little or no physical distinction between wanted sound and noise—for example, in an open-planned office, the sound of speech in a conversation between two people is wanted by them, but it may be noise to a third party who is interrupted by it.

The acoustical characteristics of spaces and of materials are usually frequency-dependent, and it is often necessary to determine the spectral composition of both wanted and unwanted sounds in order to control them. Although the meteorological effects which make predictions of outdoor sound propagation difficult are absent inside buildings, the fact that the wavelengths of low- to medium-frequency sounds are similar to, or larger than, the dimensions of rooms and the objects inside them makes the precise prediction of sound propagation inside rooms very difficult too.

In this chapter, some of the important characteristics of speech and music will be described, and the principles of design of auditoria for speech and music, or *room acoustics*, will be outlined. A most important criterion for auditoria is that no unwanted sound should be heard inside—that is sound from external sources, and from sources in other parts of the building, including mechanical services, should be inaudible. The techniques used to ensure this are similar for all types of building and the theory of airborne and impact sound transmission through buildings, its measurement and assessment and the control of building services noise and vibration are discussed in Chapter 5.

4.2 CHARACTERISTICS OF SPEECH

Although the scientific analysis of speech was stimulated in the early part of the 20th Century, with the advent of telephone and radio, and much of this work is still valid, a great deal of research in this field has been carried out more recently, including that needed specifically to develop techniques for automated speech production and speech recognition by computers.

Human speech is produced when the vocal chords are used to modify the flow of exhaled air from the lungs. Rapid vibration of the chords transforms the flow into a series of puffs, which are heard as a buzz, the frequency of which is dependent on the rate of vibration of the vocal chords. This buzz is then modified by the *vocal tract*, which consists of the throat, nose and mouth, including the tongue and lips. The shape of the lips plays a part in the production of speech sounds and so closely are the lip shapes allied to specific sounds that many deaf people are able to understand conversation reasonably well from the visual clues provided by lip movements alone. Even for normal hearing people, the visual clues are useful in aiding recognition of speech sounds. Any lip configuration can theoretically be used with any tongue configuration, but a language has normal combinations that any native speaker finds difficult to contradict—this sometimes affects the ability of a person to pronounce some speech sounds of a foreign language correctly.

The basic sounds of any language are called *phonemes*, which are units of significant sound in a given language. For example, the words 'heat' and 'heap' each consist of three phonemes, two of which are the same and one (the final phoneme in this case) is different: by changing this third phoneme, the meaning of the word changes. There are about thirty-eight phonemes in English, many of which are common to other languages. The *vowel* sounds usually have fairly definite spectral characteristics and contain the major part of the energy available in connected speech. The *consonants* tend to have short duration and less energy, although they contain more information. When speech sounds are spoken in connected phrases, or sentences, the individual sounds are

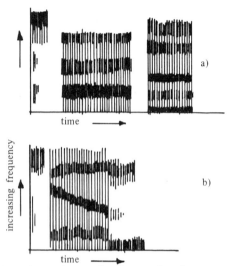

Fig. 4.1. Modification of speech sounds when spoken in connected phrases; a) phonemes uttered separately, b) phonemes spoken in connected speech.

modified, which make it difficult to determine objective rules for automatic speech recognition (Fig. 4.1).

Fortunately, it is not necessary for listeners to perceive each phoneme correctly in connected speech, provided that they are reasonably familiar with the language and the topic. There are many redundancies present in connected speech, which allow the meaning to be comprehended even if some of the phonemes are not heard. However, if words or syllables are spoken in isolation, it is much more difficult to detect them correctly—it is well known that only a limited, carefully chosen vocabulary should be used in circumstances where correct word recognition is essential, in air traffic control communication, for example. On the other hand, it is possible for a listener with good hearing ability to understand connected speech in the presence of high noise levels, or among other competing voices, and this makes the provision of *speech privacy* in open planned offices extremely difficult. Although much of the research concerning *speech intelligibility* has assumed that listeners have normal hearing and are familiar with the language, in a typical population there will be a number of people with hearing disabilities and also, frequently, some to whom the language is not familiar.[4.1, 4.2] Such people are disadvantaged when listening to speech under noisy conditions. Young children, also, have a more limited vocabulary than adults and there is some evidence to show that they need better acoustical conditions compared to adults, for the same degree of intelligibility.[4.3]

The acoustic power of human speech is quite limited, the average long-term sound power for connected speech is about 34 μwatts for men and about 18 μwatts for women. However, the instantaneous speech power ranges from about 0.01 μwatts for the quietest voice uttering the quietest sound up to about 5 000 μwatts for the loudest voice uttering the loudest sound. The long-term average dynamic range of conversational speech is about 30 dB, including peaks which are about 12 dB higher than the average levels. The speech frequency range is from about 200 Hz to 6300 Hz, with the most important components being in the range from about 1000 to 4000 Hz. This is one reason why people who have suffered loss of hearing due to excessive noise exposure have such difficulty in understanding speech, since this type of hearing loss is usually greatest around the 4000 Hz band.

One method of representing the relative importance of different frequency bands to speech intelligibility is the *dot field*. Fig. 4.2 shows the typical range of male speech levels in a normally furnished room of about 10 m² area.[4.4] If the noise level in a room such as this is plotted on the dot field, the ratio of the number of dots above the noise level to the total number of dots gives the Articulation Index, AI. This in turn may be related to the possibility of understanding connected speech. 'Good' Speech Intelligibility occurs when the AI is over about 0.45; 'Excellent' Speech Intelligibility requires the AI to be over 0.65. On the other hand, 'good' Speech Privacy is said to exist if the AI is less than 0.1 and 'confidential' Speech Privacy is obtained if it is less than 0.05.[4.5] This is a simplified method of assessment of speech communication conditions: other factors, particularly the reverberation time in the room affect intelligibility, since reverberant or time-delayed reflected sounds act as a masking noise, preventing perception of the next phonemes in connected speech.

Intelligibility is only one of the important characteristics of speech as used in theatres, etc. Dramatic use of the voice to express emotion is very important, and this may involve wider dynamic ranges than in normal conversational speech. As mentioned earlier, the acoustic power available in the human voice is strictly limited, and thus the size of an auditorium or theatre in which voice levels will be sufficiently

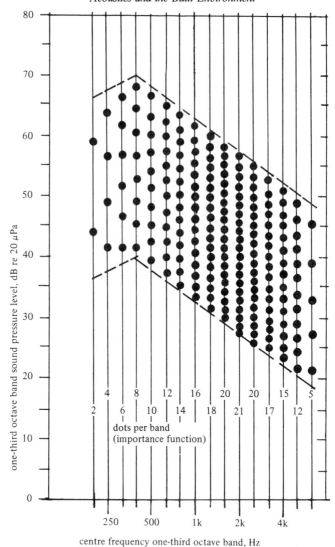

Fig. 4.2. The relative importance of speech frequency bands for intelligibility, as represented by the 'Dot Field'.

high over the whole audience area is also limited. If larger audiences are to be accommodated then it is necessary to include some form of electronic voice reinforcement, at least for the more distant parts of the audience. See Section 6.8.10.

4.3 CHARACTERISTICS OF MUSIC

It is difficult to derive objective criteria for the design of concert halls and other auditoria for music, since both value judgements and emotions are involved in people's reactions. Considerable advances have been made over the last few decades with respect to understanding people's perceptions of the complex signals produced by musical instruments in rooms, and the theory of room acoustics design has also been studied intensively. However, it still does not appear to be possible to determine precisely whether a new concert hall will receive critical acclaim, and there have been major acoustical problems even in some recently completed buildings.

Music probably began when primitive people became conscious of the sound made by striking a hollow log or a stretched animal skin. These sounds were then made rhythmically, perhaps copying the rhythm of heartbeats or breathing. After speech developed, people began to sing, and from this developed melody, or the singing of a succession of sounds of different *pitch*. For many years, rhythm and melody were the only elements of music, and later, harmony, or the playing or singing of several notes together, developed. The early emphasis on melody led to the development of different methods of organising changes of pitch, which is analogous to changing sound frequencies. Although, theoretically, any of the continuum of frequencies could be used in music, when instruments with fixed tuning are used, such as a piano for example, it is necessary to limit the choice to certain discrete frequencies.

Pythagoras carried out some of the earliest recorded investigations of musical sounds. He was interested in the ratios of the lengths of vibrating strings which gave rise to 'pleasant' intervals of sound (under a constant tension). He found that this occurred every time the length was halved, which is equivalent to a doubling of the sound's frequency. The name given to this basic interval of Western music ($2f/f$) is the octave. (The frequency ratio of two successive tones which are subjectively evaluated as a correct musical octave is slightly greater than two, however.) Other 'pleasant' frequency ratios were also determined and complete musical scales were developed.

A great variety of musical instruments now exists in many parts of the world, some of which are unique to particular musical traditions. In the West, recorders were in secular use in the 12th century and flutes were developed two centuries later. Of the many types of early stringed instruments, the bowed viol family gained supremacy and it was developed to accompany the different registers of the human voice. Early orchestras had only a few players, and this is still true of so-called 'chamber music'. However, a modern symphony orchestra may consist of over one hundred players and many different instruments, the latter being broadly grouped into strings, woodwind, brass and tympani. Most musical instruments are designed to produce fundamental frequencies (lowest frequency components) corresponding to the requirements of musical scales, together with a series of harmonics, which are higher frequencies bearing a simple relationship to the frequency of the fundamental. The instruments are also usually designed to discriminate against the production of other sounds. In simple terms, the fundamental frequency determines the pitch of the sound and the harmonics affect the musical quality, or the characteristic *timbre*. According to the requirements of the music being played, the sounds are produced at a rate of from 15 to 20 sounds per second, and as each sound has a duration of from 100 ms to 2000 ms or more there is a considerable temporal overlap. The sound produced by each instrument has an *onset* time, a *steady-state* period and a *decay;* during the initial onset of the sound, the fundamental and then the various harmonics become established, and considerable variations in instantaneous spectra may occur: these variations are called *transients*. There is evidence to suggest that an ability to hear these initial transients of a sound enables a listener to distinguish one instrument from another. It is therefore important that each member of the audience receives sufficient direct sound from each instrument, without masking by reflected sound. It is also necessary to provide many reflected sounds in order to achieve the full musical effect.

One problem that has resulted from the large size of orchestras required to play the classical music repertoire is that the concert halls needed to accommodate the audiences necessary for economic viability are also very large, and it is difficult to provide excellent listening conditions in such spaces. Fig. 4.3 shows the approximate dynamic and frequency range encompassed by a classical orchestra.

4.4 SOUND PROPAGATION IN ROOMS

4.4.1 Direct and Reverberant Sound Fields

In Section 1.2 the propagation of energy from a point source of sound in a free field was described, and it was shown that the sound pressure

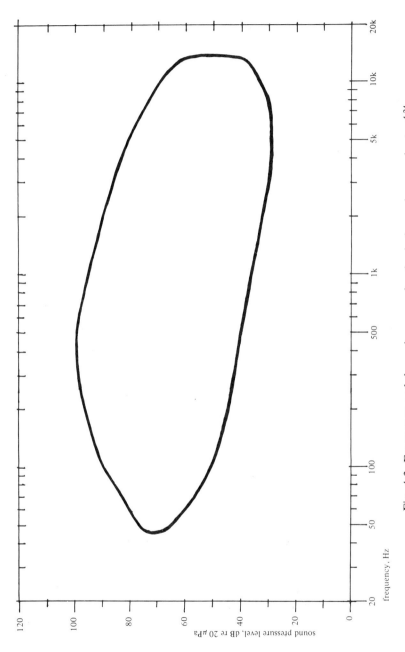

Fig. 4.3. Frequency and dynamic range of a classical symphony orchestra. [4.24]

level was reduced by 6 dB for each doubling of distance from the source. However, if the sound source is enclosed in a normal room, there will be many reflected sounds from the walls, floor, ceiling and objects in the room. A room is said to be perfectly *diffuse* if the sound waves travel in all directions with equal probability and the sound energy density, E, (Equation 1.7) is the same everywhere. In practice, it is difficult to build a room which will support a perfectly diffuse sound field, although so-called *reverberation* chambers, which are used for acoustical testing, are designed for this purpose. In practical rooms, provided that the source is far removed from any surfaces, there is usually a small 'free field' area near the source, where the sound travelling directly from the source to the receiver will predominate and where the sound will attenuate approximately according to geometrical spreading.[4.6] In this area, the energy density due to the direct sound may be found from:

$$E_D = (WQ_\theta)/(4\pi r^2 c) \hspace{2cm} [4.1]$$

where

E_D = Energy density due to the direct sound, at a distance r from the source in a direction θ, J/m^3

W = source power, watts

Q_θ = source directivity factor, direction θ, (see Sect. 1.6.3)

c = sound velocity in air, m/s

Further from the source, the *reverberant sound field* is predominant. This sound field is composed of all the sound waves that have been reflected one or more times from the room's surfaces. If the room is reasonably diffuse, half of the sound waves will have vector components from right to left and half from left to right. If a small area ΔS is located in the reverberant field, it will receive energy from one-half of these sound waves and those incident normally on the surface will each contribute a sound power of ($I\Delta S$), and those incident at an angle θ will contribute ($I\Delta S\cos\theta$), where I is the intensity of each wave; thus effectively only one-quarter of the total sound power falls on the area ΔS (See Fig. 4.4). If this area is located on one of the room's surfaces, some of the energy will be absorbed, because the impedance Z_S of the surface will usually differ from the impedance of air, Z_{air}. (See Section 1.2). The amount of energy absorbed depends on the material's *sound absorption coefficient*, α, which is defined as the ratio of the sound energy absorbed by the surface to the airborne sound energy incident upon it. The value of α varies with the angle of incidence of the sound wave and the so-called statistical sound absorption coefficient, α_{stat}, is averaged

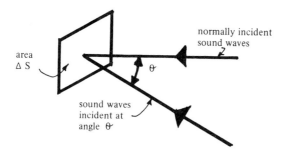

area
Δ S

normally incident
sound waves

θ

sound waves
incident at
angle θ

Fig. 4.4. Sound incident on a small surface in a room.

over all possible angles of incidence, which can only occur in a perfectly diffuse sound field. The total power absorbed by the small surface, ΔS, is:

$$[(E_R c)/4] \times [\Delta S \alpha_{stat}] \text{ watts} \qquad [4.2]$$

where

E_R = the energy density of the reverberant sound field, J/m^3
c = the velocity of sound in air, m/s
ΔS = the area of small surface, m^2
α_{stat} = the sound absorption coefficient of the surface material

The total power lost from the reverberant field is obtained by summing Equation 4.2 over all the room's surfaces; alternatively the *average sound absorption coefficient* of all the surfaces, $\alpha_{stat,\,av}$, may be used:

$$\alpha_{stat,av} = (\alpha_{stat,1}S_1 + \alpha_{stat,2}S_2 + \ldots \alpha_{stat,n}S_n)/\Sigma S \qquad [4.3]$$

where

$\alpha_{stat,1}\alpha_{stat,2}, \alpha_n$ = absorption coefficients of the different surface materials

S_1, S_2, S_n = areas of the different surface materials, m^2, (ft^2)

ΣS = total area of all the surfaces, m^2, (ft^2)

The total sound power lost from the reverberant field is then

$$[(E_R c)/4] \times [\Sigma S \alpha_{stat,av}] \qquad [4.4]$$

After a sound source has been emitting energy for a short period of time, the sound level in the room reaches a *steady state*, i.e. it reaches a constant level. This implies that in the steady state the power supplied to the reverberant field must be equal to the power remaining after the first reflection. Thus in the steady-state:

$$W(1 - \alpha_{av}) = [(E_R c)/4] \times [\Sigma S \alpha_{stat,av}]$$ [4.5]

where W = the sound power of the source, watts, thus

$$E_R = [(4W)/(c \Sigma S \alpha_{stat,av})] \times [1 - \alpha_{stat,av}]$$ [4.6]

The quantity $(\Sigma S \alpha_{stat,av})/(1 - \alpha_{stat,\ av})$ is known as the *Room Constant* R, thus Equation 4.6 may be written as:

$$E_R = (4W)/(cR)$$ [4.7]

The total sound energy density in the room, in the steady state, i.e. with a constant sound source operating, is the sum of the direct and the reverberant sound energy density,

$$E = E_D + E_R$$ [4.8]

This may be written in terms of sound pressure as

$$|p^2| = W \rho c \left[\frac{Q_\theta}{4 \pi r^2} + \frac{4}{R} \right]$$ [4.9]

and in the quantity usually measured, sound pressure level, in decibels, as

$$L_p = L_W + 10 \log \left[\frac{Q_\theta}{4 \pi r^2} + \frac{4}{R} \right]$$ [4.10]

However α_{stat} is difficult to measure, and the data usually available relates to the Sabine absorption coefficient, α, determined from reverberation room measurements (see Appendix A). The value of α is usually larger than that of α_{stat} by an amount similar to that by which R is larger than $\Sigma S \alpha_{stat,av}$; thus Equation 4.10 may be approximated by:

$$L_p = L_W + 10 \log \left[\frac{Q_\theta}{4 \pi r^2} + \frac{4}{\Sigma S \alpha_{av}} \right]$$ [4.11]

The usefulness of the relationship given in Equation 4.11 is that the relative importance of the direct and reverberant sound field may be calculated for reception points at different distances from a sound source. If the reception point is in the direct sound field, any reduction in the reverberant sound field achieved by increasing the absorption coefficients of the room's surfaces will have no effect on the overall sound level at that location. Even if the reception point is in the reverberant sound field, increasing the room's absorption may not have a great effect on reducing the total sound pressure level. An example of this type of calculation is given in Section 6.7

4.4.2 Sound Build-up and Decay

4.4.2.1 Reverberation time

In theatres, concert halls and auditoria, the steady-state situation, as described above, is rarely of importance, since the wanted sound is continually changing. Each phoneme of speech, each note of music is emitted, propagated in the room and then reduced to inaudibility. For speech perception, the most important aspects are the propagation and decay of each sound; however, for music, as discussed in Section 4.3 the build-up process is also very important, as is the sequence of reflections received at listening positions. The decay process is called reverberation, and the *reverberation time* of a room is defined as the time taken for a sound to decay by 60 decibels after the source is stopped. Reverberation time, T_{60}, is related to the size of the room and its total absorption. W.C.Sabine was the first to quantify reverberation time,[4.7] and he derived an empirical equation:

$$T_{60} = KV/A \qquad [4.12]$$

where

T_{60} = the reverberation time, sec
K = a constant (0.161 SI units, 0.049 British units)
V = the volume of the room, m^3, ft^3
A = the total absorption in the room, m^2, ft^2 ($= \Sigma S \alpha_{av}$)

Since the absorption coefficient of a material is frequency dependent, T_{60} will also vary with frequency unless a careful selection of materials with different sound absorption characteristics is made.

Eyring and others derived theoretical relationships for the decay of sound in a room.[4.8] It is assumed that the sound field in a room is perfectly diffuse, i.e. the sound energy density, E, is uniform and the sound waves are travelling in all directions with equal probability. It is further assumed that the surfaces of the room are uniformly absorbent and that their absorption may be characterised by the mean sound absorption coefficient, α_{av}. The number of times per second that a given sound wave will be incident upon a room boundary will depend on the average distance travelled between the boundaries. This distance is the Mean Free Path and it can be shown to be equal to $4V/\Sigma S$ for spaces that are reasonably regular geometrically. Thus the number of reflections occurring in one second will be $c/(4V/\Sigma S)$, where c is the speed of sound in air. After each reflection the remaining energy will be $(1 - \alpha_{av})$ less, and the average intensity, I, after n reflections will be:

$$I = I_0(1 - \alpha_{av})^n \tag{4.13}$$

where I_0 = the initial intensity, W/m^2
Thus after 1 second,

$$I = I_0(1 - \alpha_{av})^{c/(4V/\Sigma S)} \tag{4.14}$$

If this is expressed in decibels, the change in intensity is $10 \log I/I_0$ and the decay rate, D, becomes:

$$D = 10 \log \frac{I_0}{I} \text{ dB per second} \tag{4.15}$$

$$D = \frac{10c}{4V/\Sigma S} \log \left(\frac{1}{1 - \alpha_{av}} \right) \tag{4.16}$$

Since T_{60} is the time for the signal to decay by 60 dB,

$$T_{60} = \frac{60}{\dfrac{10c}{4V/\Sigma S} \log \left(\dfrac{1}{1 - \alpha_{av}} \right)} \tag{4.17}$$

Simplifying, this results in the Eyring equations:

$$T_{60} = \frac{0 \cdot 07V}{-S\log(1 - \alpha_{av})} \quad \text{(in SI units)} \tag{4.18}$$

$$T_{60} = \frac{0 \cdot 0212V}{-S\log(1 - \alpha_{av})} \quad \text{(in British units)} \tag{4.19}$$

If natural logarithms are used, rather than logarithms to the base 10, the Eyring equation is similar to the original empirical Sabine equation. The latter becomes progressively more inaccurate as the average absorption coefficient increases, predicting longer values for T_{60} than will occur. However, in many practical cases, the additional 'accuracy' achieved when using the Eyring equation is not really achieved since several of the assumptions made in the theoretical derivation are often not justified in practical situations. For example, a truly diffuse sound field is very difficult to achieve in a real room, as discussed earlier; secondly, there is usually considerable variation in the absorption characteristics of the various room boundary materials and contents (including people), which means that the assumption that an equal amount of energy will be absorbed each time a wave is reflected is incorrect. For general design purposes, except in the case of studios or similar rooms, the Sabine equation is sufficiently accurate.

Criteria for reverberation times for auditoria are well-established, and because of the relative ease both in estimating this quantity at the design stage, and of measuring it in the completed building, it has become a significant part of the auditorium design process. In general, rooms to be used primarily for speech should have short reverberation times, to avoid masking of the following phonemes; for music, longer reverberation times are generally required, in order that the audience may perceive the musicians' aesthetic intentions. Longer reverberation times are expected and therefore acceptable in large rooms than in smaller ones. The reverberation time should be reasonably constant over the frequency range at least from 500 to 4000 Hz; in rooms for music T_{60} may be longer in the lower frequencies, up to a maximum of about 150% in the lowest frequencies. Fig. 4.5 gives some recommended mid-frequency values of T_{60} for rooms of different sizes used for different purposes.

Often it is necessary to design an auditorium to be suitable for both speech and music. There are two approaches in this case: one method is to establish the priority purposes of the auditorium and to design for the first priority, with perhaps some compromise for the secondary ones; the alternative is to provide for variation in the reverberation time, according to its usage at any given time. The latter provision, although it may be costly, has the advantage that compromise criteria are not required. Methods of achieving variable reverberation time in an auditorium are several: firstly (since T_{60} is dependent on volume) it may be possible to provide means of altering the size of the auditorium, by opening or closing off balconies, stage house areas, etc.; secondly, rotating or sliding panels of absorbent/reflective materials may be used to change the average absorption coefficient; and thirdly, electronic methods of varying reverberation may be installed. This last method is becoming much more sophisticated and thus more acceptable to critical audiences; it also has the advantage that it may be possible to install such systems in completed buildings if requirements change.

4.4.2.2 Room modes

If a sound source emits a single frequency, it will propagate through the air as a sinusoid. If the source is located between two parallel walls, and a sound ray (the normal to the wavefront) is incident normally on one of the walls and no phase change takes place on reflection, the reflected energy will be in phase with the incident energy. Thus a compression

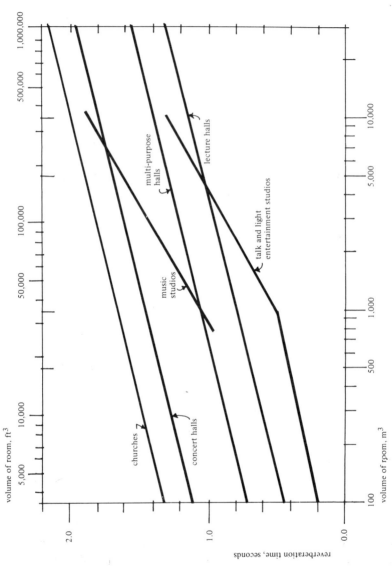

Fig. 4.5. Recommended mid-frequency reverberation times for auditoria used for different purposes (studios after Gilford[4.25]).

will coincide with a compression and a rarefaction will coincide with a rarefaction, etc. At the same time the sound ray travelling from the source in the opposite direction will also be normally incident on the opposite wall and it will also be reflected in phase. If the distance between the two parallel walls is a simple multiple of the wavelength of the sound, all the reflections will be in phase and a *standing wave* will be set up. The result of this is that there will be very large variations in sound level across the room as there will be maximum particle displacements corresponding to the coinciding maxima of compressions and rarefactions (at half-wavelength distances apart), called *antinodes*, and zero displacements in between, called *nodes*. (See Fig. 4.6). Frequencies at which this phenomenon occur are called *room modes*.

The frequencies at which modes will occur in a regular room having plane, parallel walls, floor and ceiling may be estimated from:

$$f_{p,q,r} = \frac{c}{2} \sqrt{\left(\frac{p}{L}\right)^2 + \left(\frac{q}{W}\right)^2 + \left(\frac{r}{H}\right)^2} \qquad [4.20]$$

where

$f_{p,q,r}$ = the p, q, r, mode, Hz
c = speed of sound in air, m/s (ft/s)
p, q, r = 0 or any integer
L, W, H = length, width and height of room, m (ft)

If two of the values of p, q, r are zero, then the standing wave will be one-dimensional; if one value is zero, the wave will be two-dimensional, etc. If a room has a cubic shape, then modes will occur at the same frequency in all three dimensions. Room modes are obviously undesirable in any room, and particularly unwanted in auditoria. Although modes cannot be avoided entirely, careful selection of room dimensions, avoiding simple multiples of height, width and length, for example, will ensure that they are distributed as evenly as possible. As the frequency of the signal increases there are increasing numbers of modes excited, which tend to interfere with each other, and the sound pressure becomes more uniform throughout the room. Widely spaced modes are most often a problem at low frequencies, and particularly in small rooms such as studios. In such spaces it is necessary to estimate the frequencies at which the modes will occur and to incorporate special selective sound

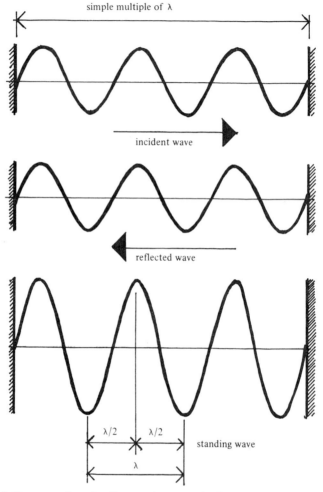

Fig. 4.6. Room modes: the distance between the boundaries is equal to simple
multiples of the wavelength of the sound.

absorbent systems to minimise their effect. (See Section 4.6.4). According to Kuttruff,[4.9] the number of modes, N_f, from 0 to a selected upper frequency limit, f, can be estimated from:

$$N_f = \frac{4\pi V}{3}\left(\frac{f}{c}\right)^3 + \frac{\pi S}{4}\left(\frac{f}{c}\right)^2 + \frac{Lf}{8c} \qquad [4.21]$$

where

V = the volume of the room, m^3 (ft^3)
c = speed of sound in air, m/s (ft/s)
S = total area of all the surfaces of the room, m^2 (ft^2)
L = total length of all the edges of the room, m (ft)

4.4.2.3 Other room acoustics criteria

Reverberation time is a rather crude measure of the acoustical quality of an auditorium: it usually varies little over the whole audience area, and yet it is well known that the perceived sound is different near the front of a concert hall than towards the rear. This is because the temporal sequence of the direct and reflected sounds and their spectral balance usually changes markedly from one location to another. A major study of important concert halls and auditoria was conducted by Beranek and published in 1962.[4.10] He proposed a number of subjective criteria and their physical correlates. Some of these are listed, together with an assessment of their current importance, in Table 4.1.[4.11]

Several major concert halls were constructed in the 1950s and 1960s according to 'modern' acoustic design theories. Some of them, including the Royal Festival Hall in London and the Philharmonic Hall in New York, were severely criticized and they were compared unfavourably with older halls, particularly with the typical 19th Century 'shoe-box' type hall such as Boston Symphony Hall and the Musikvereinsaal in Vienna. This stimulated further research into concert hall design, and a number of additional criteria were proposed. The new halls differ physically in many respects from those built in the 19th Century, they tend to be much larger than the old ones for example, partly because they accommodate larger audiences and partly because of higher standards of comfort and safety. The new halls also provide good sight lines for the majority of the audience, which was not the case in many of the older ones, and their cross-sections tended to be quite different. (See Fig. 4.7 a) and b)). Marshall investigated reflection sequences in two idealized hall shapes, one modelled on the classical shoe-box and the other a wide, rectangular hall with a relatively low ceiling, typical of more recently built halls.[4.12] He predicted that in a wide hall, all reflections from the side walls would be masked by reflections from the ceiling, but this would not be the case in the shoe-box hall: he suggested that in the wide hall 'Spatial Responsiveness' would be lacking.

Jordan observed that during a musical performance the total decay

TABLE 4.1[a]
Beranek's Criteria for Concert Hall Acoustics

Attribute	Objective expression	Ideal value	Current importance
1. Intimacy	Initial Time Delay Gap between receipt of direct and first reflected sound	10–20 ms	Still thought to be important, but direction of reflected sounds should be included
2. Liveness	Average of reverberation time, T_{60}, at 500 & 1 000 Hz	1·9 s	General variation of reverberation time with frequency should be controlled
3. Warmth	Ratio of low- to mid-frequency reverberation time, T_{60}: $(T_{125} + T_{250})/\{2[(T_{500} + T_{1\,000})/2]\}$	1·2–1·25	As for 2
4. Loudness of direct sound	Distance to centre of main floor seats	20 m ($\cong$ 60 ft)	Important, in addition, clear sightlines are necessary
5. Loudness of reverberant sound	Ratio of mid-frequency reverberant time, T_{60} to volume of hall		More detailed analysis of early reflection sequence necessary
6. Diffusion	Wall and ceiling irregularities	Adequate	More detailed calculation of surface modelling necessary
7. Balance and blend	Design of performers' end of hall, response hall as perceived by performers	Good	Very important
8. Ensemble	Performers' ability to hear each other	Easy	Very important

[a] From Ref. 4.11.

process over 60 decibels is rarely heard. He proposed the Early Decay Time, EDT, as a criterion.[4.13] This is determined by extrapolating the slope of the first part of the reverberation decay curve, between 0 and -10dB. (See Fig. 4.8). It is certainly more relevant than the conventional T_{60} if the decay rate is not constant. However, the value of EDT does not vary greatly from one part of an auditorium to another, and its accurate calculation at the design stage is difficult.

A major criterion for listening to music in a concert hall is that the audience should perceive the difference between being present at a live performance and hearing the sound via loudspeakers at home. It has long been recognized that a true 'stereo' effect cannot be achieved by playing the same recorded sound signal through two separated loudspeakers, since a 'phantom' sound source from which all the sound appears to originate is formed mid-way between the two loudspeakers. Even if two different signals are played through the two loudspeakers, the effect is not the same as being present in an auditorium. Schroeder and others have found that for the most preferred listening conditions in an auditorium it is necessary for the sound signals received by the left ear to be *incoherent* with those received by the right ear.[4.14] Schroeder quantified this effect as Binaural Similarity, defined as the peak of the correlation function of the first 80 ms of an impulse response, with an interaural delay range of 1 ms. Ideally, the value of Binaural Similarity should be zero. It is physically related to the directional distribution of the reflected sounds. Reflected sounds coming from a plane ceiling will produce rather coherent signals at a listener's two ears, whereas reflections from lateral directions, e.g. from side walls, should not. However, the problem of coherent ceiling reflections may be overcome if specially modelled diffusing surfaces are used.[4.15]

Other criteria have been suggested, such as Blauert's Auditory Spaciousness,[4.16] Houtgast and Steeneken's Modulation Transfer Function,[4.17] and Plenge et al.'s 'Horsamkeit' (Hearingness).[4.18] These latter authors proposed a modification of existing criteria to take into account the physical extensiveness of orchestral sound sources. One important aspect that is sometimes overlooked was called by Jordan the 'Musicians Criterion'.[4.13] If musicians do not hear sound from other parts of the orchestra, and if they do not receive some indication of the response of the room to their own instruments, they will not be able to perform to their best ability.

4.5 GEOMETRICAL DESIGN OF AUDITORIA

It is evident that the detailed sequence and spectra of the direct and reflected sounds received by the two ears of each listener is critical in determining subjective response. It is not yet possible to specify exactly which sequences will comprise 'excellent' acoustics. Some designers have decided to continue using the traditional shapes for opera houses, concert halls and drama theatres, which will support traditional sound fields

a)

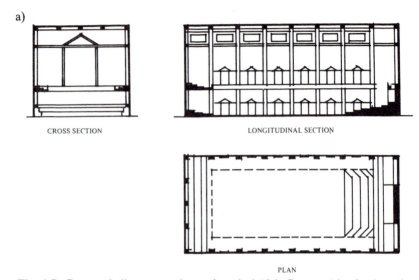

CROSS SECTION   LONGITUDINAL SECTION

PLAN

Fig. 4.7. Concert hall cross-sections, a) typical 19th Century 'shoebox' seating approximately 1660 people; width 21 m (69 ft), furthest seat 41 m (134 ft), b) mid-20th Century hall seating about 2680 people, width 36 m (118 ft), furthest seat 33 m (108 ft).

which will be recognized by their audiences (and which will inevitably include their inherent limitations). Others have developed new forms with resulting 'new' sounds—these auditoria are subjected to much critical judgement and as discussed earlier, there are frequently differences of opinion regarding their acoustic merit.

One factor that does appear to be very important is that each member of the audience should have a clear line-of-sight to the performers, and since human eyes and ears are at approximately the same level, a clear sight-line means a clear direct 'sound-line' as well. Haas[4.19] found that if a listener receives sound signals from a number of different locations, all of the sound appears to originate from the direction of the first signal received: this is known as the Haas, or precedence effect. (The direct sound will always be received before the naturally reflected sound in an auditorium, but in order to preserve the directional illusion that the signal is coming from the source on the stage if loudspeakers are used for amplification, it is necessary to locate the loudspeakers so that the distance between the loudspeaker and the listener is greater than that between the live source and the listener. If this is not possible, electronic

b)

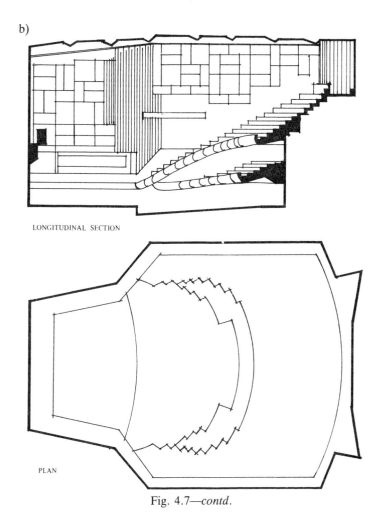

LONGITUDINAL SECTION

PLAN

Fig. 4.7—*contd.*

time delays should be incorporated in the system in order to achieve the same effect. See Section 6.8.10).

In addition to the importance of the direct sound in determining the perception of source locations, it is also the signal that has been least affected by possible selective absorption of sound frequencies at the room's boundaries. A geometrical method of ensuring good sight-lines, and thus good direct sound, is shown in Fig. 4.9. Firstly, the audience

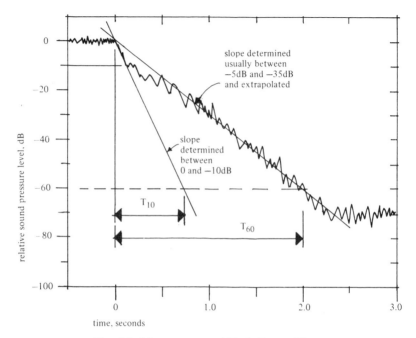

Fig. 4.8. Measurement of Early Decay Time.

seating layout should be determined; the row-to-row spacing is usually governed by regulations concerning evacuation in case of emergency, and it also depends on whether or not intermediate aisles are provided. Next, the 'sight-point' that each member of the audience should be able to see is selected—for drama theatres at least, this is preferably the front lower edge of the stage or platform. A line is drawn from the sight point to the top of the heads of the first row of the audience and extended to the second row; the heads in this row should be elevated above this line by the required clearance (a minimum of 75 mm or 3 in). This process is then repeated until the last audience row is reached. If there are balconies, a similar procedure is used. The required floor slopes (or stepping) are then determined from the respective seated head heights (approximately 1120 mm or 3 ft 8 in above the floor). It will be found that a parabolic slope is ideal, which may need to be slightly modified for practical constructions. However, any modification should be upwards, not downwards, if the sight-lines are to be maintained.

Because of the finite time taken for sound to travel around a room it is sometimes possible to detect one or more specific echoes. This is

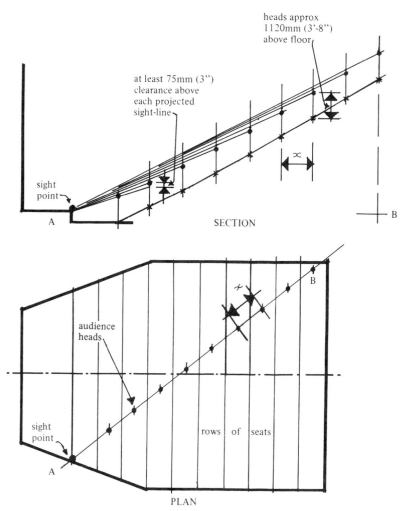

heads approx
1120mm (3'-8")
above floor

at least 75mm (3")
clearance above
each projected
sight-line

sight
point

A   SECTION   B

audience
heads

sight
point

rows of seats

A   B

PLAN

Fig. 4.9. Floor slope design for audience seating to ensure good sight-lines.

obviously undesirable, and experiments have been performed to quantify the conditions under which echoes are perceived. Haas also found that sounds that arrive more than 50 ms (50/1000 sec) after the direct sound may be perceived as distinct echoes if they have retained sufficient energy.[4.19] Sounds that arrive within 30 ms of the direct sound cannot be perceived separately, although a change in quality may be heard. Thus it is useful to determine the geometry of the auditorium by choosing the shape so that short-delayed reflected sounds are directed to the audience, particularly to those seated further from the source. Long-delayed reflections which may be heard as echoes may be avoided by calculating the relative path-lengths of direct and reflected waves. Since the speed of sound in air is approximately 344 m/s (1120 ft/s) a delay time of less than 30 ms corresponds to a path-length difference of less than about 10 m (34 ft) and a delay time of over 50 ms corresponds to a path-length difference of over 17 m (56 ft). If it is not possible to avoid long-delayed reflected sounds the possibility of echoes being perceived may be reduced by covering the surfaces from which they are reflected with sound absorbent materials.

The height of the ceiling should be chosen to provide sufficient volume to develop the required reverberation time for the auditorium's purpose (see Section 6.8). The shape of the ceiling (and of the walls) may be determined geometrically—either manually or by computerised graphics. Fig. 4.10 shows a simple method of geometrical construction which will ensure that useful reflected sound is directed to the audience from the room's surfaces. The method commences with the selection of a suitable point 'A', which may be the underside of the proscenium arch, for example. A line is drawn from the source position (the speaker, or actor, for example) to point A and another line from A to the first row of auditors for whom reflected sound is desired (in this case, the fifth row). The angle between these two lines is then bisected, which gives the normal to the direction of the required surface slope, S_1 (since the angle of reflection must be equal to the angle of incidence). If S_1 is extended towards the direction of the source, the Image, I_1 of the source in the plane S_1 may be found by extending the normal from the source to the plane an equal distance beyond the plane. All sound originating at the source and reflected by S_1 will appear to come from the image source, I_1. In order to determine how far S_1 should extend, draw a line from I_1 to the last row of the audience, which gives the cut-off point, B. The exercise is now repeated, with a line from the source to B and another from B to, say, row 7 in the audience; this angle is bisected and the surface slope S_2 is determined. Image source I_2 and point C are determined in the same way as before. This procedure may be repeated as often as necessary. The cross-section of the room should also be examined in a similar fashion.

Obviously, there are other considerations in the determination of ceiling shapes in theatres, in particular, for example, the location of lighting bridges, structural supports, etc. may intrude on the desired

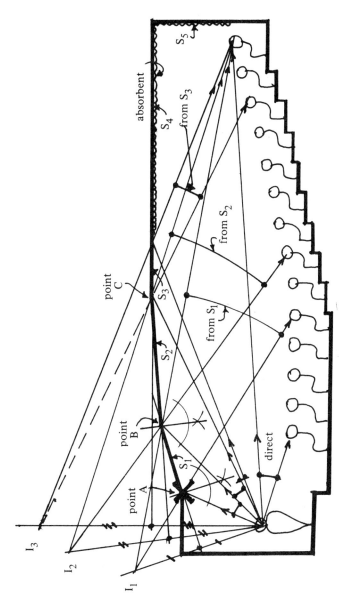

Fig. 4.10. Geometrical design of auditoria to ensure good early reflected sound is directed to the audience.

reflecting planes, but the geometrical method may be adapted to allow for such modifications.

A similar procedure may be used to determine useful side wall shapes for reflected sounds. It will also identify wall and ceiling areas that may need to be covered with sound absorbent materials to avoid echoes.

It should always be remembered that manual methods of determining the shape of an auditorium are severely limited in their validity. For example, only two dimensions are examined at a time, thus three-dimensional reflections are not included; secondly, the optical analogy on which the method relies is only valid when the surfaces are large compared with the wavelength of the incident sound. Thus low-frequency, long wavelength sounds are not included in the analysis.

Several computer programmes are now available for the geometrical analysis of auditoria shape, some of which are three-dimensional and thus overcome one of the limitations of the manual method. However, any method based on the optical analogy is also only valid for mid- to high-frequency sound components.

4.6 SOUND ABSORBENT MATERIALS

4.6.1 Introduction
In Sections 4.4.1 and 4.4.2 it was shown that the absorptive properties of the boundaries of a room have an important effect on the sound field developed. The audience and seating make a major contribution to the total absorption present and in addition, the absorption by the air in the room may be significant in large rooms at high frequencies (see Section 2.4.1). It is important to realize that *every* material and system will interact with sound waves in the room, not only those selected specifically for their 'acoustic' properties. For most materials and systems, the interactions are strongly dependent on the frequency of the incident sound, and a basic understanding of the mechanisms involved is helpful in selecting suitable finishes. There are three basic types of sound absorbents: 1) porous absorbents, 2) membrane absorbent, and 3) resonant absorbents.

4.6.2 Porous Absorbents
If an airborne sound wave is incident upon a porous surface, the vibrating air particles are restricted by the walls of the pores. The particles in

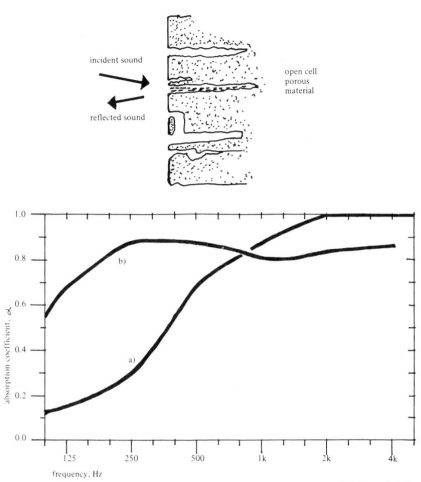

Fig. 4.11. Sound absorption characteristics of porous materials: a) 25 mm (1 in) mineral wool mounted against solid backing, b) 50 mm (2 in) mineral wool mounted over 180 mm (7 in) airspace.

the centre of each pore may respond freely to the compressions and rarefactions of the sound wave, but the air near the pore walls does not move. A shear force is thus developed, and due to the viscosity of air, some sound energy is converted into heat which is absorbed by the pore walls. In order for this absorption to take place it is obviously necessary for the vibrating air to be able to enter the pores of the surface material: a measure of this is the *flow resistance* of the material:

$$R = \Delta p/\Delta du \qquad\qquad [4.22]$$

where

R = flow resistance, rayls/m
Δp = sound pressure level difference across sample thickness, Pa
Δd = thickness of sample, m
u = particle velocity through sample, m/s

The amplitude of the air particle vibration is progressively damped by friction against the pore walls, which act as an acoustical resistance. If the value of R is too high, reflection will occur at the surface, and absorption will be lowered; if it is too low, however, the wave will pass through the material with little attenuation. The particle velocity of a sound wave is highest at $\lambda/4$ locations, and it is when the particle velocity is greatest that maximum absorption of sound energy can take place. This means that for commonly used thicknesses of porous materials (such as mineral wool or glass fibre) in buildings, say 25–50 mm (1–2 in), the absorption will be greatest at high frequencies and there will be little absorption at low frequencies. If it is necessary to achieve high sound absorption using this type of material it has to be very thick, of the order of several metres (feet). An alternative method of obtaining better low and mid-frequency absorption using a porous material is to suspend it freely at about $\lambda/4$ distance from a rigid boundary; if the flow resistance, R, of the material is correctly selected almost 100% absorption of sound energy may be obtained for the corresponding narrow frequency band.

As discussed earlier, it is usual to describe the absorption of sound by materials in terms of their absorption coefficients. These coefficients are determined by measurement, either in a reverberation chamber or in an impedance tube. (See Appendix A). If the data is to be applied in practice it is important that the relevant mounting conditions are chosen for comparison; e.g. if the data is for 25 mm (1 in) of mineral wool of a certain density, mounted 50 mm (2 in) from a rigid surface, this should not be used for other thicknesses, densities or distances from the surface. Fig. 4.11 shows some typical examples of sound absorption coefficient versus frequency for porous materials.

4.6.3 Membrane (Panel) Absorbents

If a relatively thin, airtight material is fixed at some distance from a rigid surface, it will act as a mass-spring system and exhibit characteristic resonances (preferred frequencies of vibration). When a sound wave impinges it will tend to force the system into vibration; if the frequency of the sound corresponds to one of the system's resonant frequencies there will be a maximum transfer of energy. Since the thin panel has inertia and is also damped due to fixing at its edges, there will be some sound energy converted into mechanical energy, and thus 'absorbed'. However, since the panel is itself set into vibration it will re-radiate energy back into the room: thus its efficiency as an absorbent is limited.

Practical panel absorbents tend to be most effective in the low frequency range, and there is a marked maximum absorption at the resonant frequency; generally, as the surface density of the panel and/or the depth of the air space between the panel and the rigid surface increase, the frequency of maximum absorption decreases. If it is required to broaden the effective absorbent frequency range it is necessary to place some porous absorbent material in the air space.

An important corollary of panel absorption is that if an auditorium is lined with large areas of thin panels, these will act as low frequency sound absorbents, leading to a loss of low frequency components in music. If, for aesthetic reasons, large areas of timber panelling are required, it is essential that the panels are backed with a rigid material, such as gypsum plaster. It is also advisable to mount the panels at different distances from the rigid surfaces and also to construct them of different sizes and thicknesses. If this is done the amount of energy absorbed will be reduced, and the panels' resonant frequencies, at which the absorbent maxima occur, will differ. Fig. 4.12 shows some typical sound absorbent coefficients of a panel absorbent.

4.6.4 Resonator Absorbents

It is well known that if air is blown across the open neck of an empty container, a sound will be heard, the container acting as a resonator. There is some evidence to suggest that ceramic pots and jars with openings to the air (which act as resonators) were built under seats in ancient Greek amphitheatres and into some walls of medieval churches for acoustic purposes.

Simple resonators are named after Helmholtz who first described their characteristics. A Helmholtz resonator consists of a volume of contained air connected to the air in the room through a constricted neck and

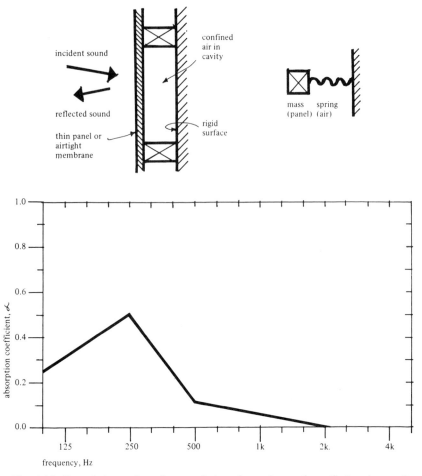

Fig. 4.12. Sound absorption characteristics of membrane (panel) absorbents: 3 to 5 mm (1/8–3/16 in) pannelling over 25 mm (1 in) airspace.

opening. The impinging sound energy causes the air in the neck to vibrate, and because of its constricted volume, this air acts as a mass supported on a spring (the air in the main chamber). As for any mass-spring system, there will be a resonant frequency at which there will be maximum vibration. If a sound of this frequency impinges on the res-onator there will be a maximum transfer of energy, and absorption will also be greatest at this frequency because of frictional losses.

Helmholtz resonators are most effective over a narrow frequency range, normally at low frequencies. They can be finely tuned and they are very useful in small broadcast studios, for example, when it is necessary to reduce the effect of widely spaced room modes (see Section 4.4.2.2). The approximate resonant frequency for a resonator having a circular opening is given by:

$$f_{res} = \frac{c}{2\pi} \sqrt{\frac{S}{l'V}} \quad \text{Hz} \qquad [4.23]$$

where

f_{res} = frequency of maximum absorption, Hz
c = velocity of sound in air, m/s (ft/s)
S = surface area of the neck, m^2 (ft^2)
l' = effective length of the neck = $(l + \pi r/2)$, m (ft)
r = radius of opening, m (ft)
V = volume of contained air in chamber, m^3 (ft^3)

The maximum absorption, at resonance, of a single undamped resonator depends on its resonant frequency[4.20] and may be estimated from:

$$A = 0.159 \, (c/f_{res})^2 \qquad [4.24]$$

where

A = total equivalent absorption area, m^2 (ft^2)
c = velocity of sound in air, m/s (ft/s)
f_{res} = resonant frequency, Hz

If it is wished to broaden the response of this type of resonator, a porous absorbent material may be placed in the chamber—this will have the effect of reducing the maximum absorption at resonance. As in the case of membrane absorbents, some sound energy will be re-radiated into the room from the Helmholtz resonator. Fig. 4.13 shows an example of the absorption characteristics of a Helmholtz resonator.

4.6.5 Perforated Panel Absorbents

Helmholtz resonators are expensive to construct and they are only effective over a narrow frequency band. A similar mechanism of absorption occurs with perforated panels mounted at some distance from a rigid surface, whereby the perforations in the panel act as a series of 'necks' sharing the same 'chamber'. The air in the necks acts as a series of masses supported by a spring (the air in the space between the

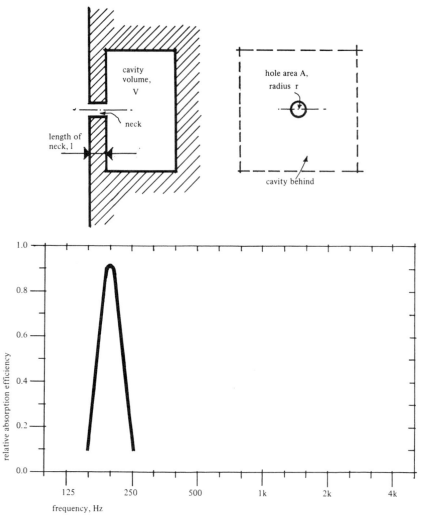

Fig. 4.13. Sound absorption characteristics of Helmholtz resonators (undamped).

perforated panel and the rigid surface behind), and there is a resonant frequency at which sound absorption will be a maximum. The resonant frequency of a given system may be estimated from:

$$f_{res} = \frac{c}{2\pi} \sqrt{\frac{P}{dl'}} \quad Hz \qquad [4.25]$$

where

f_{res} = frequency of maximum absorption, Hz
c = velocity of sound in air, m/s (ft/s)
P = perforation ratio (hole area/panel area)
d = distance of panel from the wall, m (ft)
l' = effective thickness of neck $= (1 + \pi r/2)$, m (ft)
r = radius of the hole, m (ft)

As with individual resonators, placing a porous absorbent material in the air space will broaden the effective absorbent range, whilst reducing the maximum absorption at resonance; it tends to raise the resonant frequency compared to the empty cavity situation. Frequently a thin membrane or tissue is placed between the porous absorbent in the cavity and the panel; this affects the flow resistance of the porous material, and thus the overall absorption characteristics of the system. Perforated panel absorbents tend to be most effective in the mid-frequency range. Again, it is important that relevant system's absorption data are selected in any practical applications, since changing the thickness of the panel, the perforation ratio and the depth of the air space, as well as placing porous absorbents in the cavity, will have a marked effect on performance at different frequencies. Fig. 4.14 shows some typical absorption coefficients for two perforated panel systems.

In some applications, perforated panels are provided with slits rather than with holes. This complicates the calculation of the resonant frequency, since this is somewhat dependent on the frequency of the impinging sound wave.

4.6.6 Audience and Seat Absorption

In a typical concert hall, the audience itself is the major absorbent of sound energy, at least in the medium to high frequencies. Unfortunately, it is difficult to predict how much sound will be absorbed by the audience, as it is greatly affected by its configuration. For example, if the audience is well spaced out and the seats are raked steeply from the performance end of the theatre, each person will be exposed to sound energy from many directions; alternatively, if the audience is packed closer together, and the seating is on a flat floor, each person will receive much less incident sound energy.

In calculations of reverberation time, audience absorption is either treated on a per capita basis as X m² (X ft²) equivalent absorption area

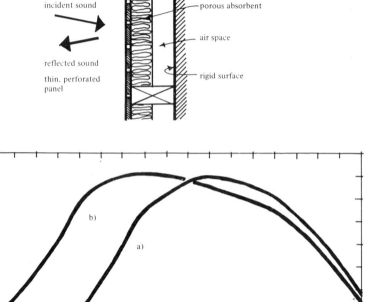

Fig. 4.14. Sound absorption characteristics of perforated panel absorbents: a) thin panel perforated 10% over 25 mm (1 in) mineral wool over 25 mm (1 in) airspace, b) thin panel perforated 10% over 50 mm (2 in) mineral wool over 25 mm (1 in) airspace.

per person, or as a surface area having absorption coefficients as for other surfaces in the room. If the per capita basis is chosen it is necessary to add the equivalent absorption area of each unoccupied seat. Wherever possible, the seats should be chosen to have equivalent absorption to a person, so that the reverberation time is independent of the number of people actually present in the room. However, this is not always possible,

TABLE 4.2
Equivalent Absorption Area Per Capita of a Selected Audience

	Centre frequency of octave band, Hz						
	63	125	250	500	1 000	2 000	4 000
Equivalent absorption area, m^2	0·14	0·19	0·28	0·33	0·35	0·37	0·39
Equivalent absorption area, ft^2	1·5	2·0	3·0	3·5	3·8	4·0	4·2

TABLE 4.3
Equivalent Absorption Area of an Audience Related to Total Seating Area

	Centre frequency of octave band, Hz						
	63	125	250	500	1 000	2 000	4 000
Equivalent absorption area, m^{2}ft^{2a}	0·34	0·52	0·68	0·85	0·97	0·93	0·85	0·80

[a] If audience area measured in m^2, equivalent absorption area is in m^2.
If audience area measured in ft^2, equivalent absorption area is in ft^2.

particularly in multi-purpose auditoria where much, if not all, of the seating is required to be stackable.

A guide to the absorption of an audience on a per capita basis is given in Table 4.2. These figures relate to normal audience densities with a moderately raked and medium-upholstered seating. If the audience is less densely packed and if the seating is heavily raked and/or heavily upholstered, the figures should be increased by up to about 15%.

Beranek[4.21] proposed the use of absorption coefficients, to be applied to an augmented *audience area* calculated by adding to the actual audience seating area, the area of the aisles within and around the audience, up to a maximum of 1.06 m (3.5 ft) width each, plus any area used as standing room. The coefficients he proposed, to be applied to traditionally shaped concert halls and opera houses, are given in Table 4.3.

These figures should be used with caution, and it is possible that in any given audience seating configuration, large deviations will be found in practice. Lower absorption will occur if the seating is not upholstered; large groups of children, for example in school auditoriums, are also less absorbent than adults, and the amount of clothing worn will also affect the results. Since the audience has such a major effect on total absorption, it is essential that an audience is present during the final tuning of any important new auditorium and that provisions have been made for adjustment of the absorption of other surfaces.

An important characteristic of sound propagation across audiences was described by Schultz and Watters, and Sessler and West.[4.22, 4.23] They found excess attenuation of low-frequency sound over the first 12 to 15 rows of seats from the stage. It occurred principally in the frequency range from about 120 and 200 Hz, and could amount to as much as 20 dB in excess of inverse square law attenuation. The frequency of maximum attenuation was found to be related chiefly to the height of the seats; it corresponds to a seat height of approximately $\lambda/4$ and it is attributed to wave interference effects. If the seats are not raked from the front of the auditorium, it is evident that there will be excessive absorption of low-frequency sound, leading to complaints of 'lack of bass sound'.

4.7 SUMMARY

In this chapter the essential characteristics of sound propagation within rooms have been described. The effects of room shape, surfacing materials and the audience on the sound field have been examined, and criteria for rooms for listening to speech and music have been discussed. In the case of auditoria designed for speech the main requirements are good direct sound, low reverberation time and sufficient loudness. For music there should be good direct sound, a number of reflected sounds following the direct sound received within 30 ms, the signals at the two ears should preferably be uncorrelated, and there should not be any perceptible echoes; the reverberation time should be designed in accordance with recommendations for the size of the auditorium. Specific examples of different types of auditoria will be considered in Chapter 6.

Although guidance is provided for the geometrical design of auditoria and for the selection of their surfacing materials, there are still many difficulties inherent in specifying and then designing an auditorium to support the required acoustical characteristics and the construction of physical acoustic models may be necessary for important auditoria such as concert halls. Computer models are also useful, particularly if they take into account the three-dimensional nature of the space. One advantage of a physical acoustical model is that it is possible to 'listen' to the auditorium before it is built, and to determine subjectively its acoustical quality. (See Section 6.8.1.)

REFERENCES

4.1. Suter, A.H. Speech recognition in noise by individuals with mild hearing impairments. *J.Acoust.Soc.Amer.* 78, 1985, pp 887-900.

4.2. Florentine, M. Non-native listeners' perception of American-English in noise. *Internoise* 85, Munich, 1985, pp 1021-1024.

4.3. Neuman, A.C. & Hochberg, I. Children's perception of speech in reverberation. *J.Acoust.Soc.Amer.* 73, 1983, pp 2145-2149.

4.4. Cavanaugh, W.H., Farrell, W.R., Hirtle, P.W. & Watters, B.G. Speech privacy in buildings. *J.Acoust.Soc.Amer.* 34, 1962, pp 475-492.

4.5.—*Acoustics—methods of assessing and predicting speech privacy and speech intelligibility.* AS 2811-1985, Standards Association of Australia.

4.6. Embleton, T.F.W. Sound in large rooms, Chap.9, *Noise and Vibration Control*, L.L. Beranek, Ed. McGraw Hill, 1971.

4.7. Sabine, W.C. *Collected Papers on Acoustics.* Harvard Univ. Press, Cambridge, USA, 1922, pp 3- 68.

4.8. Eyring, C.F. Reverberation time in 'dead' rooms. *J.Acoust.Soc.Amer.* 1, 1929, pp 217-241.

4.9. Kuttruff, H. *Room Acoustics*, App.Sc.Pubs.London, 1973, p 56.

4.10. Beranek, L.L. *Music Acoustics and Architecture*, Wiley, NY, 1962.

4.11. Lawrence, Anita. Sightlines and soundlines—the design of an audience seating area. *App.Acoust.* 16, 1983, pp 427-440.

4.12. Marshall, A.H. A note on the importance of room cross section in concert halls. *J.Sound Vib.* 5,1967, pp 100-112.

4.13. Jordan, V.L. Auditoria acoustics: Developments in recent years. *App. Acoust.*8,1975, pp 217-235.

4.14. Schroeder, M.R., Gottlob, D. & Siebrasse, K.F. Comparative studies of European concert halls; correlation of subjective preference, with geometric and acoustic parameters. *J.Acoust. Soc.Amer.* 56, 1974, pp 1195-1201.

4.15. Schroeder, M.R. Binaural dissimilarity and optimum ceilings for concert halls: More lateral sound diffusion. *J.Acoust.Soc.Amer.* 65, 1979, pp 958-963.

4.16. Blauert, J. & Lindemann, W. Explorative studies on auditory spaciousness. *Proc. Acoustics and Theatre Planning for the Performing Arts*, Vancouver, 1986. (12th Int.Cong. on Acoustics Satellite Symposium) pp 39-44.

4.17. Houtgast, T. & Steeneken, H.J.M. The Modulation Transfer Function in room acoustics as a predictor of speech intelligibility. *Acustica* 28, 1972, pp 66-73.

4.18. Plenge, G., Lehmann, P., Wettschureck, R. & Wilkens, H. New methods in architectural investigations to evaluate the acoustic qualities of concert halls. *J.Acoust.Soc.Amer.* 57, 1975, pp 1292-1299.

4.19. Haas, H. Uber den Einfluss einers Einfachechos auf die Horsamkeit von Sprache. *Acustica* 1, 1961, pp 49-58.

4.20. Zwikker, C. & Kosten, C.W. *Sound Absorbing Materials*, Elsevier, Amsterdam, 1949, p 135.
4.21. Beranek, L.L. *Music, Acoustics & Architecture*, Wiley , N.Y. 1962, p 543.
4.22. Schultz, T.J. & Watters, B.G. Propagation of sound across audience seating. *J.Acoust.Soc.Amer.* 36, 1964, pp 885-896.
4.23. Sessler, G.M. & West, J.E. Sound transmission over theatre seats. *J.Acoust. Soc.Amer.* 36, 1964, pp 1725-1732.
4.24. Knudsen, V.O. & Harris, C.M. Acoustical designing in architecture. Wiley, N.Y.1950, p 21.
4.25. Gilford, C. Acoustics for radio and television studios. Peter Peregrinus, London, 1972, p 216.

CHAPTER 5

Sound Transmission in Buildings

5.1 AIRBORNE SOUND TRANSMISSION

5.1.1 Airborne Sound Transmission through Homogeneous Panels

In Section 1.2 it was shown that when a sound wave travelling through air is incident upon a boundary, such as the ground, or a wall, the ratio between reflected and absorbed (transmitted) energy depends on the impedance of the material comprising the boundary relative to that of air. If a wall divides two rooms, for example, a proportion of the airborne sound energy incident on the wall's surface in Room 1 will be reflected back into Room 1 and a proportion will be 'absorbed' by the wall: the 'absorbed' energy will then travel through the solid material of the wall and be incident on the wall's boundary with the air in Room 2. Again, a proportion of this energy will be 'absorbed' by the air in Room 2, and some will be reflected back into the wall (See Fig. 5.1). (Some sound energy will be truly absorbed, that is it will be converted into heat energy by doing mechanical work.) Thus the principle factor determining the amount of sound transmitted from the air in Room 1 to the wall and then from the wall to Room 2, and thus the amount of sound transmitted by this path from Room 1 to Room 2 is the ratio of the impedance of air to that of the material of the dividing wall. As stated earlier (Section 1.2) the impedance of a medium, $Z = \rho c$, where ρ is the density of the medium and c is the velocity of sound energy within it, the latter itself being dependent on the elasticity and density of the medium; it follows that the greater the density of the material, compared to that of air, the larger will be the impedance ratio and the less sound will be transmitted between the rooms. This is known as the 'Mass-Law' theory of sound transmission. It can be shown that the expected reduction of sound level

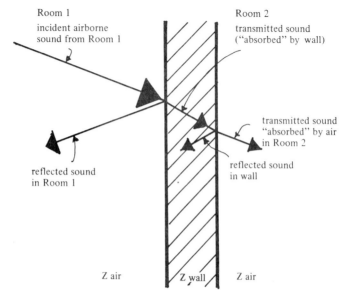

Fig. 5.1. Transmission of airborne sound through infinite homogeneous panels; reflection at impedance boundaries.

between two rooms separated by a partition, or the *sound transmission loss* of the partition R′, in decibels, is given by:

$$R' = 20 \log \left[(\omega M)/(2 Z_{air}) \right] - 5 \qquad [5.1]$$

where

    ω = the angular frequency of the incident sound ($= 2\pi f$, Hz)
    M = the surface density of the partition, kg/m² (lb/ft²)
    Z_{air} = the characteristic impedance of air

This equation may be simplified to:

$$R' = 20 \log (fM) + K \qquad [5.2]$$

where

    f = the frequency of the incident sound, Hz
    M = the surface density of the partition, kg/m² (lb/ft²)
    K = a constant = −47 dB, M in kg/m² (= −34 dB, M in lb/ft²)

Equation 5.2 shows, theoretically, that the sound transmission loss will increase by 6 decibels for each doubling of surface density (or, for a given material, each time the thickness is doubled), and by 6 dB for each doubling of frequency (i.e. for each octave). However, the attenuation predicted using this equation should be regarded as the maximum sound transmission loss between rooms that could be attained for randomly incident sound; in practice, the attenuation achieved is usually several decibels lower.

The above theory strictly applies only to impedance boundaries of infinite extent, and it assumes that each part of the solid medium reacts independently. In practice, these assumptions are only valid over a limited range of frequencies (generally in the medium frequency range). In air, or in any gaseous medium, sound may only be propagated in the form of longitudinal waves, but in a solid material, shear and bending, or flexural waves, may also be present. In a plate, the longitudinal wave velocity is given by:[5.1]

$$c_{L'} = \sqrt{\frac{E}{\rho_p(1 - \sigma^2)}}$$ [5.3]

where

$c_{L'}$ = longitudinal wave velocity, m/s (ft/s)
E = Young's modulus of the material, N/m² (lb$_f$/ft²)
ρ_p = density of the material, kg/m³ (0.03 lb$_m$/ft³)
σ = Poisson's ratio ($\cong 0.3$ in most cases)

In practice, a simplified expression (strictly applicable to thin bars where the lateral extension is small) may be used to estimate longitudinal wave velocity:

$$c_L = \sqrt{\frac{E}{\rho_\rho}}$$ [5.4]

The bending, or flexural wave velocity in the plate is given by:

$$c_B = \sqrt{1.8\,d\,f\,c_{L'}}$$ [5.5]

where

c_B = the bending wave velocity, m/s (ft/s)
d = thickness of the plate, m (ft)
$c_{L'}$ = the longitudinal wave velocity, m/s (ft/s)

Equations 5.4 and 5.5 show that longitudinal sound waves of all frequencies travel at the same velocity, but flexural waves do not. This means that in a solid, the wave-form of a complex signal is not preserved and the medium is said to be *dispersive*.

Real panels in buildings are finite in size and they are able to support many *resonances*, depending on their dimensions and on the longitudinal wave velocity. If an incident sound wave coincides with a resonant frequency of the panel, good sound transmission will occur, since the compressions will coincide with compressions, rarefactions with rarefactions, etc. If the panel is a wall between two rooms, for example, at resonant frequencies the sound attenuation between one room and another will depend primarily on damping at the panel edges. The resonant frequencies of a thin rectangular plate, supported but not clamped on its edges, are:

$$f_{n,m} = 0.45 \, c_L \, h \left[\left(\frac{n_x}{l_x} \right)^2 + \left(\frac{n_y}{l_y} \right)^2 \right] \qquad [5.6]$$

where

$f_{n,m}$ = the n, m mode of resonance, Hz
c_L = the longitudinal wave velocity, m/s (ft/s)
h = the thickness of the panel, m (ft)
n, m = any integers
l_x, l_y = the panel dimensions, m (ft)

The lowest resonant frequency occurs when n, m = 1. Below the lowest resonant frequency, the transmission loss is said to be *stiffness controlled*, that is, mass and damping are unimportant. In practical cases, panel resonance affects sound transmission in the lower frequency range. At a frequency of about twice that of the lowest resonant frequency, Mass-Law sound transmission, as described above, occurs. This relationship continues until the *critical frequency* is reached. This is the lowest frequency capable of exciting the coincidence effect. As shown by Equation 5.5 the bending, or flexural wave velocity, is dependent on frequency; at a certain frequency and angle of incidence of the airborne sound wave it will have the same projected wavelength as the bending wave in the panel at the same frequency. This has a similar effect as a resonance, and the sound of this frequency will be transmitted readily from one room to the other. The critical frequency may be estimated from:

$$f_c = \frac{c^2}{1.8\,h\,c_{L'}} \quad Hz \qquad [5.7]$$

where

f_c = the critical frequency, Hz
c = the velocity of sound in air, m/s (ft/s)
h = the thickness of the panel, m (ft)
$c_{L'}$ = the longitudinal wave velocity in the material, m/s (ft/s)

This may approximated as:

$$f_c \cong \frac{c^2}{1\cdot8h} \sqrt{\frac{\rho_p}{E}} \quad Hz \qquad [5.8]$$

where

ρ_p = the density of the material, kg/m^3 (lb/ft^3)
E = Young's modulus, N/m^2 (lb$_f$/ft^2)

Fig. 5.2 shows a typical curve of sound transmission loss against frequency for a panel exhibiting low frequency resonances, mid-frequency Mass-Law behaviour and coincidence.

For thick, dense partitions, such as masonry walls, the value of f_c is low and the coincidence effect occurs in a similar frequency range to panel resonances and is thus not readily differentiated from them. However, for panels that are relatively thin, such as window glass, plywood, plasterboard, etc. the coincidence effect is noticeable and it occurs in an important part of the frequency range, typically around 1000 to 4000 Hz (see Fig. 5.3).

5.1.2 Airborne Sound Transmission through Multiple Skin Partitions

Frequently two panels are built close to each other, e.g. a cavity brick wall or a double glazed window. In such cases the overall sound attenuation might be expected to be double that of a single wall or window; however, this does not occur because of coupling effects. In the Mass-Law region, it is true that attenuation is greatly increased, because the sound energy encounters four impedance boundaries rather than the two in a single panel. Theoretically, in this region, the sound transmission

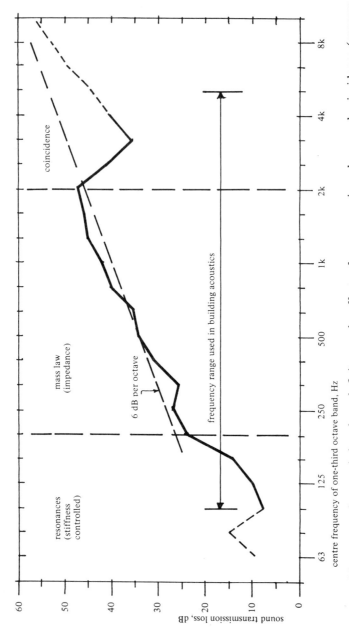

Fig. 5.2. Airborne sound transmission through finite panels; effects of resonance, impedance and coincidence (gypsum boarding on a stud framework).

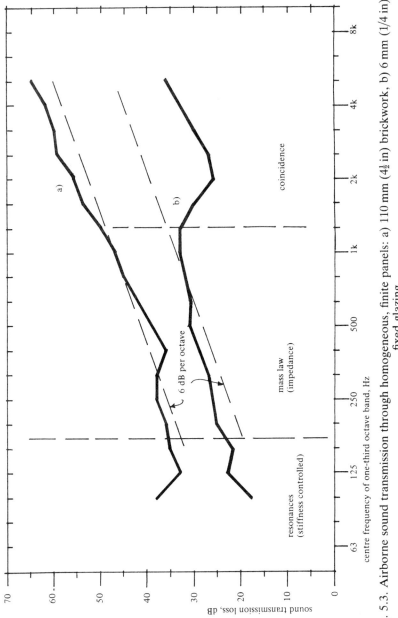

Fig. 5.3. Airborne sound transmission through homogeneous, finite panels: a) 110 mm (4½ in) brickwork, b) 6 mm (1/4 in) fixed glazing.

loss will increase by 12 dB per octave. However, the air in the cavity forms an elastic coupling between the two skins and at low frequencies a mass-air-mass resonance occurs. The approximate resonant frequency may be estimated from:[5.2, 5.3]

$$f_0 = F \sqrt{\frac{M_1 + M_2}{M_1 M_2 d}}$$ [5.9]

where

f_0 = mass-air-mass resonance frequency, Hz
M_1, M_2 = surface density of two skins, kg/m^2 (lb$_m$/ft^2)
d = thickness of air space, m (ft).
$F = (1/2\pi)(\rho c^2)^{1/2} = 60$, d in m, M_i in kg/m^2
($\cong 50$, d in ft, M_i in lb/ft^2)

In the case of typical double glazed windows, f_0 occurs in the lower frequency range and results in a pronounced dip in the curve of sound transmission loss versus frequency.

For frequencies above $c/2d$ (where the cavity width, d, is equal to one half the wavelength) standing wave patterns, or cavity resonances may form in all three dimensions of the cavity space; low frequency cavity resonances depend on the overall dimensions of the panels (width and height) and the high frequency ones on the cavity width. Theoretically, in this range, the transmission loss should be:

$$R' = R_1' + R_2' + 10 \log (A/S)$$ [5.10]

where

R' = the overall sound transmission loss, dB
R_1', R_2' = the sound transmission losses of panels 1 and 2, dB
A = the equivalent absorption in the cavity, m^2 (ft^2)
S = the area of the panel, m^2 (ft^2)

If it is assumed that the panel surfaces are not absorbent, as is the case for a double glazed window system, the equivalent absorption A, in Equation 5.10, depends on the provision of absorbent linings around the perimeter of the cavity. A may be determined from Equation 5.11:

$$A = \alpha d(2h + 2w)$$ [5.11]

where

α = absorption coefficient of perimeter surfaces
d = cavity depth, m (ft)

h, w = height and width of panels

These equations show that where it is possible to line the cavity, reveals with sound absorbent material, the effect of cavity resonances will be reduced and attenuation will be improved. An approximate guide to the theoretical maximum sound transmission loss of a double-skin construction, at the mid-range frequency of 500 Hz, was given by Ford and Lord:[5.4]

$$R' = 20 \log (Md) + K \hspace{2cm} [5.12]$$

where

R' = average sound transmission loss, 500 Hz
$M = M_1 + M_2$, kg/m² (lb/ft²)
d = width of cavity, m (inches)
K = constant, = 34, metric (= 16, imperial units)

This equation predicts an increase of 6 dB for each doubling of total surface density and also an increase of 6 dB for each doubling of cavity depth. However, Quirt[5.2, 5.3] found in practice that for double glazed windows an increase of only about 3 dB was obtained for a doubling of d. A slope of 6 dB per octave (not 12 dB as would be expected from simple Mass-Law theory) may be assumed above 500 Hz up to the critical frequency, f_c, and from 500 Hz down to the air-mass-air resonance frequency, f_0. If the two panels are of the same material and the same thickness, then f_c will be at the same frequency for both of them; a smoothing of the sound transmission loss characteristic will be obtained if the surface density differs between the two panels, since they will then have different f_c values (See Fig. 5.4). Generally, practical constructions are found to have much lower sound transmission loss values than would be predicted theoretically, and it is always advisable to use actual measured results for the construction to be used.

5.1.3 Statistical Energy Analysis of Sound Transmission Loss

Another theoretical treatment of sound transmission between rooms through single and double panels is called Statistical Energy Analysis, or SEA.[5.5] The method includes the interactions between the panel and the air in the rooms on either side of it, as well as the behaviour of sound in the panel. As shown in Section 4.4.2.2, the air in a room will support a number of modes, or resonances, and panels also have resonant frequencies, depending on their dimensions. The sound power flow

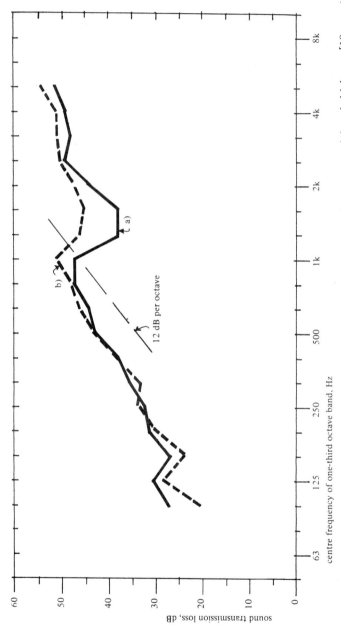

Fig. 5.4. Airborne sound transmission through double panels; a) both panels of same material and thickness [10 mm + 10 mm (3/8" +3/8") glass, 50 mm (2") airspace], b) panels of same material but different thickness [6 mm + 10 mm (1/4" + 3/8") glass, 50 mm (2") airspace.

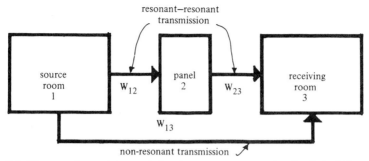

Fig. 5.5. Theory of sound transmission between rooms, using Statistical Energy Analysis.

between any two systems, e.g. from the airborne sound in the room to a panel, is assumed to be proportional to their average modal energy difference. The power flows from the system with the higher modal energy to that with the lower energy.[5.6] The actual modes present in each system are difficult to estimate, but in the SEA theory, averages are made over space, time and frequency.

Brekke presented methods for calculating sound transmission loss using SEA.[5.7] It is necessary to make separate calculations for 'resonant' and 'non-resonant' frequency transmissions. The latter are frequencies at which a panel is forced to vibrate by an impinging sound wave, the former are those which are reflected at the boundaries and which form standing waves. Fig. 5.5 illustrates the principle factors considered. In the case of a single panel between two rooms, there are three elements; energy from the resonant modes in the source room is transmitted to the resonant modes in the panel, W_{12}, and then from these modes to the resonant modes in the receiving room, W_{23}. In addition, there is forced, non-resonant transmission, from the source room to the receiving room, W_{13}. In order to use SEA analysis it is necessary to have information regarding the 'total loss factor' which includes losses at the boundaries and radiation losses, and the 'radiation factor' of the panel.

Ellmallawany compared SEA with classical methods of predicting sound transmission loss and also with experimental results. He found that SEA is generally accurate, although there are still discrepancies between theory and practice.[5.8, 5.9, 5.10, 5.11]

5.1.4 Low Frequency Sound Transmission Loss
In typical small office and domestic sized rooms, the room modes at low frequencies are widely spaced, and if the receiving room has a dominant

resonant mode the attenuation will be small or even negative. Mullholland and Lyon define low frequencies as those below which the number of modes in the room is less than 20. For a room with a volume of only 10 m³ this frequency may be as high as 500 Hz.[5.12] They considered resonant and nonresonant coupling between room modes and panels and they found that negative insulation minima occur when there is coupling between a set of nonresonant panel modes and a room mode, and that any room modes with natural frequencies less than f_{BN} will produce this effect:

$$f_{BN} = (\rho c^2 TS)/(13.8\,\pi VM) \quad \text{Hz} \qquad [5.13]$$

where

ρ = density of air, kg/m³ (lb/ft³)
c = velocity of sound in air, m/s (ft/s)
T = reverberation time of the room, s
S = area of panel, m² (ft²)
V = volume of the room m³ (ft³)
M = surface density of panel, kg/m² (lb/ft²)

They concluded that in general panel materials may be divided into two types, those for which resonant transmission has a significant effect on the sound transmission loss, such as glass, steel and aluminium, and materials for which this type of transmission is not significant, such as plasterboard, concrete and plywood. Stiffening of panels made of the second group of materials in order to reduce the effect of panel resonances will therefore not be effective. In general, low frequency variation in attenuation is caused by room resonances—this means that it is difficult to predict the practical low frequency sound transmission loss of a particular panel since it will vary according to the dimensions of the rooms between which it is located. If it is known that there will be particular low frequency components generated by sound sources in the room, then care should be taken with the room dimensions so that room modes to not correspond to these frequencies.

5.2 STRUCTURE BORNE SOUND TRANSMISSION IN BUILDINGS

Many sounds in buildings originate as impacts on solids or as induced vibration of the structure. Banging doors, footsteps and vibrating machinery are typical examples of such sources inside buildings, and ground vibrations excited by underground railways, heavy traffic or blasting may also be present. Once the sound energy has been imparted to a solid such as a wall, floor, or structural element, it will travel long distances with little attenuation—particularly if it is part of a large building. The structure borne energy will also be radiated by the building elements and it will be perceived as airborne sound in adjacent rooms. If the vibration is sufficiently strong, it may also be felt (See Section 1.4.)

As in the case of airborne sound, attenuation of structure borne sound occurs through reflections at impedance boundaries and by losses where one element is coupled to another, e.g. where a wall joins a floor. Because of the inherent redundancies and indeterminate boundary conditions in many buildings, it is extremely difficult to predict the attenuation that will occur between the source location and a receiver in another part of the building; it may also be difficult sometimes to determine the exact location of the source. Craik[5.13] has given an expression for the coupling loss factor between two elements, such as two walls:

$$\eta = (2L\tau)/(\pi Sk) \qquad [5.14]$$

where

η = coupling loss factor
L = common boundary length, m (ft)
τ = transmission coefficient = (W_i/W_t)
S = surface area of the source panel, m^2 (ft^2)
k = the wave number of the source wall = $2\pi/\lambda$
W_i = power incident on the wall boundary, watts
W_t = power transmitted at the boundary, watts

Where one panel is joined to a number of other panels, e.g. at a room corner, the total damping at the edge, due to coupling losses, is given by Craik as:

$$\eta = (2L\alpha)/(\pi Sk) \qquad [5.15]$$

where

L = the common boundary length, m (ft)
α = the absorption at the boundary = $\Sigma\tau$
$\Sigma\tau$ = the sum of the transmission coefficients
S = the surface area of the source wall
k = the wave number = $2\pi/f$

TABLE 5.1
Approximate Attenuation Per Joint, dB re 5×10^{-8} m/s

	Centre frequency octave band, Hz						
	31·5	*63*	*125*	*250*	*500*	*1 k*	*2 k*
First 2 joints	2	3	4	6	7	8	10
> 2 joints	2	2	3	4	4	4	4

The total damping, or loss factor is the sum of the internal loss factor, or damping of the panel material, and the coupling loss factors at boundaries. The total loss factor of a panel, e.g. a wall, is:

$$\eta = \frac{(2H + 2W)c\alpha_{av}}{S\pi^2 f^{1/2} f_c^{1/2}} + \eta_{int} \qquad [5.16]$$

where

H, W = height and width of panel, m (ft)
c = velocity of sound in air, m/s (ft/s)
α_{av} = the weighted average absorption at the boundaries
S = the panel area (=HW), m² (ft²)
f = frequency, Hz
f_c = critical frequency, Hz
η_{int} = internal loss factor of the material

The transmission across corners, T- and cross-joints can be estimated if the panel thickness, longitudinal wave speed, surface density, bending stiffness, and critical frequency are known for each element.

In practice, the decrease in structure-borne energy with distance from the source occurs mostly through coupling losses, although the reduction per junction is less at greater distances from the source. Tukker suggested that after about two junctions, the attenuation losses are lower per junction;[5.14] he gives the approximate vibration attenuation per joint as shown in Table 5.1. The resulting radiated airborne sound level may be estimated from:

$$L_p = L_v + 10 \log\Sigma S + 10 \log \sigma - 10 \log (A/4) \qquad [5.17]$$

where

L_p = sound pressure level, dB re 20 µPa

L_v = velocity level of the surfaces, dB re 5×10^{-8} m/s
ΣS = surface area of the surfaces, m^2 (ft^2)
σ = radiation factor
A = total equivalent absorption area in the room, m^2 (ft^2)

Below the critical frequency, f_c, σ is approximately 0.1 and above this frequency, σ is approximately 1.0. Depending on thickness, f_c is between about 70 and 200 Hz for typical building elements such as walls and floors.

5.3 VIBRATION TRANSMISSION IN BUILDINGS

5.3.1 Whole Building Vibration

A building may be considered as a spring-mass system, which may be excited into vibration by external forces, e.g. by ground vibration caused by transportation, particularly by underground railways, or blasting or by wind. The most critical elements are usually masonry walls subjected to horizontal vibrational forces. The whole-building resonant frequencies are primarily dependent on the overall building height, and range from about 10 Hz for low-rise buildings to less than 0.1 Hz for buildings of 60 storeys or more.

As discussed earlier, building elements, such as beams, floors, walls, etc. also have resonant frequencies; typically loaded steel beams will resonate at from 5 to 50 Hz, floors and slabs from 10 to 30 Hz and domestic sized ceilings at about 13 Hz. Such elements will tend to vibrate at large amplitudes if they are subjected to vibrational excitation at these frequencies.

Although complaints are sometimes received that buildings, or parts of them, have been damaged by nearby blasting, this is not expected to occur unless the peak particle velocity exceeds about 110 to 120 mm/s. Vibration levels such as these would cause severe discomfort to the building occupants. However, very tall buildings subjected to strong winds do tend to sway laterally, and this may alarm the occupants. One survey found that 75% of the occupants of a tall building had been aware of movement during strong winds, and they also reported seeing lighting fixtures sway and water moving in baths.[5.15] It is also common to find that lightweight walls, windows and floors of domestic buildings may be excited into vibration by high level airborne sound from aircraft, trains and heavy road vehicles. In turn, objects mounted on walls, such as

Acoustics and the Built Environment

crockery, etc. may rattle. This leads to increased annoyance of the occupants.

5.3.2 Vibration Isolation

The most common requirement for vibration isolation is that energy from vibrating machinery is not transmitted into the building. In some specialized applications, for vibration-sensitive instrumentation and machinery for example, it is necessary to ensure that vibration from the building is not transmitted to the instrument or machine. As discussed in Section 5.2, sound energy will be readily transmitted from one solid to another, and the total losses from reflections, internal damping, etc. are quite small. It is therefore necessary to install special vibration isolators to attenuate transmission of the vibrational energy of the machine to the building.

For a simple, undamped mass-spring system, in the steady state, the force transmissibility, E_f, which is the ratio of the transmitted force to the applied force, is found from:

$$E_f = 1/[1 - (\omega/\omega_n)^2] \hspace{2cm} [5.18]$$

where

ω = angular frequency of the exciting force, rad/s
ω_n = natural angular resonant frequency of the isolator, rad/s

A similar relationship is found for E_d, the displacement transmissibility, which is the ratio of the transmitted displacement to the applied displacement.[5.16] This shows that the amount of energy transmitted is primarily dependent on the ratio of the frequency of the exciting vibration to the natural resonant frequency of the isolator. In situations where the ratio ω/ω_n is very small, the effect of the isolator is negligible and as the ratio increases there is increased amplification; when $\omega = \omega_n$, (the exciting and resonant frequencies are the same) infinite amplification would occur, theoretically. As ω continues to increase with respect to wn, the amplification is reduced until at about $\omega = (2\omega_n)^{1/2}$ the effect of the isolator is again negligible. At frequencies of ω greater than $(2\omega_n)^{1/2}$ the transmitted force is lower than that applied, and the isolator begins to become effective. Fig. 5.6 shows this theoretical relationship between transmissibility and the frequency ratio ω/ω_n. It is obviously very important that an isolator with the correct value of ω_n is selected.

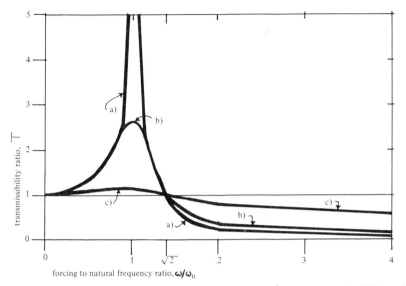

Fig. 5.6. Vibration isolation; effect of the ratio ω/ω_n on transmissibility; a) undamped, b) intermediate damping, c) critically damped. When T < 1 attenuation is effected, when T > 1, amplification occurs.

The resonant, or natural frequency, ω_n, of an undamped, single-degree-of-freedom isolator is given by:

$$\omega_n = (y/\mu)^{1/2} \qquad [5.19]$$

where

    ω_n = the natural frequency of the isolator, rad/s
    y = the spring constant, N/m
    μ = the mass, kg

This may be written as:

$$f_n = (1/2\pi)(\omega_n) \qquad [5.20]$$

where

    f_n = natural frequency of the isolator, Hz

The static deflection of an isolator, is given by:

$$\delta_{st} = \mu g/y \qquad [5.21]$$

where

    δ_{st} = static deflection, mm

g = acceleration due to gravity

By substituting in Equation 4.39,

$$f_n = A(1/\delta_{st})^{1/2} \tag{5.22}$$

where

A = constant = 15.8, δ_{st} in mm (= 3.13, δ_{st} in in)

Equation 5.22 is important since it shows that the natural frequency of a mass on a simple isolator is a function only of its static deflection, that is, its deflection under its own mass. It also implies that the designed natural frequency of an isolator will only be achieved if 1) it is correctly loaded and 2) it is able to deflect the required amount. The natural frequencies of some common isolators used in buildings are shown in Fig. 5.7.

In practice, damping occurs naturally or it is intentionally introduced; this has the effect of avoiding excessive amplification if the machine has variable speeds of vibration and the situation of $\omega = \omega_n$ cannot be avoided at all times. However, it also reduces the effectiveness of the isolator at the higher frequencies.[5.17] To measure the effectiveness of an isolating system with low damping, the transmissibility, T, is given by:

$$T = \frac{1}{(f/f_n)^2 - 1} \tag{5.23}$$

If viscous damping is present, this becomes:

$$T = \sqrt{\frac{1 + 4D^2(f/f_n)^2}{[1 - (f/f_n)]^2 + 4D^2 (f/f_n)^2}} \tag{5.24}$$

where

D = the damping ratio = C/C_0
C = actual damping
C_0 = critical damping (no movement)

The damping ratio varies from about 0.005 for a material such as steel to 0.02 for a natural rubber.

This previous discussion assumes that the supporting construction does not vibrate itself; however, with the increasing use of lightweight, less rigid buildings, the resonance of the supporting structures must also be considered. If possible, the natural frequencies of the isolator and the supporting structure should be similar to each other, but of course

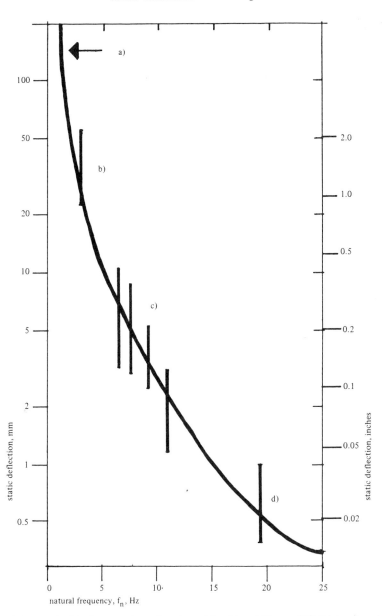

Fig. 5.7. Practical isolators. a) steel springs, b) rubber 150 mm (6″) thick, c) cork 200 − 75 mm (8″–3″) thick, d) felt 25 mm (1″) thick .

dissimilar to that of the forcing frequency. The resonant frequencies of a coupled system such as an isolator on a floor may be estimated from:

$$f_x = \sqrt{\tfrac{1}{2}(f_2^2 + f_1^2) \pm \tfrac{1}{2}\sqrt{(f_2^2 - f_1^2)^2 + 4f_1^2 f_3^2}} \qquad [5.25]$$

where

$f_1 = A\,(y_1/m_1)^{1/2}$
$f_2 = A\,[(y_1 + y_2)/m_2)]^{1/2}$
$f_3 = A\,(y_1/m_2)^{1/2}$
$A = \text{constant} = 15.8,\ \delta_{st}$ in mm ($=3.13,\ \delta_{st}$ in in)
$y_1 = $ stiffness of isolator, kg/mm
$y_2 = $ effective stiffness of supporting floor, kg/mm
$m_1 = $ mass of the plant to be isolated, kg (lb)
$m_2 = $ effective mass of supporting floor, kg (lb)

5.4 MEASUREMENT AND ASSESSMENT OF AIRBORNE SOUND ATTENUATION

5.4.1 Measurement of Airborne Sound Attenuation

There are two main applications for the measurement of airborne sound attenuation: the first is laboratory measurement to determine the sound transmission loss of specific building elements and the second is *in situ* measurements of sound transmission of elements in actual buildings. The International Standards Organization, ISO, has published a number of standard methods for measurement of sound insulation,[5.18] and many countries have also published their own standards in this area. Although there are some differences in detail, they all include the following principles.

5.4.1.1 Laboratory measurement of sound transmission loss
The definition of the Sound Transmission Loss, R, of an element is:

$$R = 10 \log{(W_s/W_r)} \qquad [5.26]$$

where

$R = $ the Sound Transmission Loss in decibels
$W_s = $ the sound power incident on the partition under test, watts
$W_r = $ the sound power transmitted through and radiated by the partition under test, watts.

Sound power cannot be measured directly, but it may be derived from measurements of sound pressure levels if the sound fields are diffuse. In this case, sound transmission loss, for diffuse incidence is expressed as:

$$R = D + 10 \log (S/A) \qquad\qquad [5.27]$$

where

D = the average sound pressure level difference, $dB = L_{ps} - L_{pr}$
L_{ps} = the average sound pressure level in the source room, dB re $20 \mu Pa$
L_{pr} = the average sound pressure level in the receiving room, dB re $20 \mu Pa$
S = the area of the partition under test, m^2 (ft^2)
A = the equivalent absorption area in the receiving room, m^2 (ft^2)

This method relies upon diffuse sound fields being set up on either side of the partition under test, which requires the use of two special reverberation rooms. These are of massive construction, structurally isolated from each other, with adjacent openings to receive the test partition of about 10 m² (100 ft²) area. They are designed to ensure that the only sound energy transmitted from one room to the other will be through the specimen under test. The specimen is then built into the aperture, and, if necessary, it is allowed to dry or to cure. The rooms must be large enough to support sufficient room modes in the lowest frequencies that will be used for the measurements, and volumes of 200 m³ (7 000 ft³) are recommended. Care is required in the selection of the room dimensions and it is usual to install special fixed or rotating-vane diffusers. Several microphone positions are used in each room in order to determine the average sound pressure levels, and the sound source is usually band-limited white or pink noise. Measurements are usually carried out in one-third octave bands from 100 or 125 Hz up to 3 200 or 4 000 Hz. The reverberation time in the receiving room is measured by interrupting the signal and obtaining the decay rates over the same range of frequencies.

Laboratory measurements such as this are time-consuming and expensive. Unfortunately, even when laboratories comply precisely with the relevant standard measurement procedures, there are some differences in the results obtained for similar specimens in different laboratories. Thus, if fine distinctions need to be made between one system and another, it is preferable that they both be tested in the same laboratory.

5.4.1.2 Field measurement of airborne sound transmission

Since, as shown above, it is not always possible to reproduce the performance of similar elements in laboratory situations, it is not surprising that there are many difficulties in attempting to measure the airborne sound transmission loss of an element installed in a building. In many cases the rooms either side are not large enough to support sufficient low frequency modes and they do not have adequate diffusion. There may also be flanking transmission present; that is, sound is transmitted from one room to the next by other paths, as well as through the element under test. If the circumstances are such that reasonably diffuse sound fields do exist on both sides of the partition, and that flanking transmission is negligible, the Field Sound Transmission Loss may be determined from:

$$\text{FTL} = D + 10 \log (S/A) \qquad [5.28]$$

where the symbols have the same meaning as in Equation 5.27. A measurement of this type is useful if an acceptance test forms part of a contract.

However, in many cases it is not possible to measure Field Transmission Loss, but it may be possible to measure the overall Noise Reduction between two spaces, where

$$\text{NR} = L_{ps} - L_{pr} \qquad [5.29]$$

where

NR = the noise reduction between two spaces, dB

In this case the proportion of sound energy transmitted by the different elements separating the spaces cannot be determined. However, it does give the effective overall attenuation of airborne sound, which usually is of most concern to the building's occupants. Measurements of NR may be carried out in one-third-octave or one-octave bandwidths as for STL, or, alternatively, the overall attenuation in dB(A) may be all that is required.

There have been several attempts to derive simple, short methods of measurement of airborne sound transmission in buildings, particularly when these are required to satisfy regulatory requirements.[5.19] Some of these methods are promising, at least for clear-cut cases of pass or fail.

5.4.2 Assessment of Airborne Sound Transmission Loss

In special cases, such as the design of concert halls, studios, etc. the detailed information obtained in measurements of STL is required.

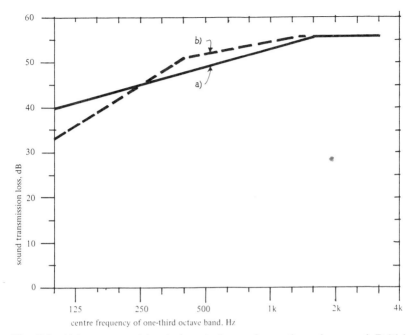

Fig. 5.8. Airborne sound insulation single number rating schemes: a) British House Grade, b) ISO 717/1.

However, for applications in dwellings or in offices, it is frequently sufficient to use a single-number rating to describe the relative performance of different types of construction. There have been two basic approaches in the development of single-number rating systems. One approach, adopted in the UK and applied to dwellings, is to determine the average sound transmission loss curve of typical forms of construction: in the case of walls, for example, the 'House Grade' is based on the STL curve of a 230 mm (9 in.) solid brick wall. The measured sound transmission loss of a particular construction, normalised to a reverberation time of 0.5 seconds in the receiving room, is then compared with this standard curve. A total adverse deviation (measured curve below standard curve) of 23 dB is allowed over the 16 one-third octave bands. (See Fig. 5.8 a). The other approach is to determine the typical spectrum of noise that is to be attenuated and to develop an ideal sound transmission loss curve that will reduce this noise to a uniform loudness level in the receiving room. This approach was used by the American Society for Testing and Materials, ASTM, in the development of the

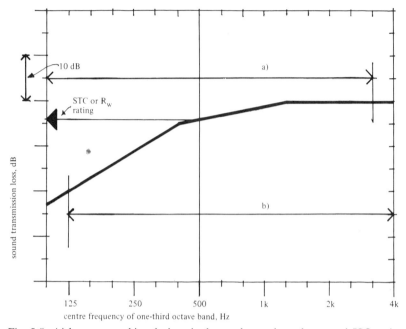

Fig. 5.9. Airborne sound insulation single number rating schemes: a) ISO rating system, b) ASTM rating system.

Sound Transmission Class, STC.[5.20] The spectrum used was averaged over spectra of typical domestic and commercial noises, which are typically flat between 250 and 1000 Hz, and which have slopes of −4 to −6 dB per octave above and below these frequencies. The shape of the ideal sound transmission loss curve derived in this manner was similar to that already standardized in Europe (Fig. 5.8 b). There is a slight difference in the frequency range used in the STC rating (125–4000 Hz) and ISO (100–3250 Hz), but for most forms of construction this is not significant (see Fig. 5.9). The measured sound transmission loss is compared with the STC (or ISO) curve and an average adverse deviation of not more than 2 dB is permitted over the sixteen, one-third octave bands. In order to avoid large discrepancies over a narrow frequency range, such as occurs in lightweight systems at coincidence, there was also a maximum deviation of 8 dB in any band. However, this restriction has now been deleted from the ISO method,[5.21] although if the deviation does exceed 8 dB in any band it should still be recorded. The rating level is the value of the 500 Hz ordinate of the rating curve when the curve-fitting rules are satisfied, and is expressed as STC (number) or as R_w or

R'_w (number) (see Fig. 5.9). In some countries, a minimum value of R'_w is required (52 dB for example), and the performance of the actual system being rated is expressed as the 'airborne sound insulation margin'—if it is positive this means that the system is better than the minimum requirement, because the reference curve can be shifted upwards; if it is negative this means that the reference curve must be shifted downwards for the curve-fitting rules to be obeyed.

Since the A-weighting curve is frequently used to describe overall noise levels as a single number, it would be logical to use this also to describe sound transmission loss or noise reduction performance. However a difficulty exists because the A-rating of a given construction depends not only on its STL properties but on the spectrum of the sound in the source room. Thus a source with prominent low-frequency components would usually result in a lower rating number than one that was mainly composed of medium to high frequencies. This could be overcome by standardizing on a source spectrum for measurement purposes, but this does not solve the problem of choosing between two different systems if the actual source to be attenuated differs markedly from that used in the measurement or rating system.

Care should always be taken when using any single number rating system—for example, it is possible that home-stereo systems with loud-speakers capable of radiating powerful low frequency signals require better low frequency attenuation performance than implied in the STC or R'_w rating systems. As mentioned previously, for critical applications it is always advisable to calculate the expected attenuation by using one-third octave band data for the source spectrum, the sound transmission loss of the construction and the reverberation time in the receiving space.

5.5 MEASUREMENT AND ASSESSMENT OF IMPACT SOUND ATTENUATION

5.5.1 Measurement of Impact Sound Attenuation

Impact sound results from the interaction between building elements and an impact source. It is therefore necessary to standardize on the source if comparisons are to be made between different elements. Although banging doors and other impacts on vertical surfaces have been found to cause annoyance in domestic situations, little progress has been made on the development of a standard method of measuring the transmission of sound energy through vertical elements. The impact test methods that have been developed are intended primarily to assess the effectiveness of a floor in attenuating footstep noise transmission to the room below. Although the International Standards Organization, ISO published standards for both laboratory and *in situ* measurement of impact sound transmission through floors many years ago, these are not universally accepted, and research aimed at developing improved methods is continuing.

5.5.1.1 Laboratory measurement of impact sound transmission

The ISO standard[5.22] requires the use of a standard ISO Tapping Machine, the construction of which is carefully specified. It has five brass or steel hammers placed in a line, the distance between the two end hammers being about 400 mm (15 3/4 in). Each hammer has an effective mass of 0.5 kg (1.1 lb) plus or minus 2.5% and an equivalent free drop of 40 mm (1.57 in) plus or minus 2.5%. The hammers strike the floor sequentially, and the time between successive impacts should be 100 ms plus or minus 5 ms.

The floor to be tested is constructed within an aperture between two structurally isolated, reverberant rooms, similar to the ones used for airborne sound transmission, but arranged vertically one above the other. The test floor should be about 10 to 20 m² (100 to 200 ft²) with the shorter edge not less than 2.3 m (7.5 ft). However, it is recommended that the size of the test floor should be similar to that to be used in a building. It is also important that the installation is similar to the actual construction, particularly with regard to joints and perimeter sealing. The tapping machine is placed in at least four different positions on the floor under test, and if the floor has beams, ribs, etc. more positions may be necessary and the row of hammers should be orientated at 45° to the direction of the beams or ribs. The resulting sound pressure level in the receiving room below is averaged over space and time, and either one- or one-third octave band filters are used over the frequency range 100 to 3150 Hz. The measured levels are corrected for absorption in the receiving room, and the normalised impact sound level is found from:

$$L_n = L_i - 10 \log (A_0/A) \qquad [5.30]$$

where

L_n = the normalised impact sound pressure level, dB re 20 μPa
L_i = the average sound pressure level in the receiving room, dB re 20 μPa

A_0 = the reference sound absorption = 10 m^2 (100 ft^2)

A = the measured sound absorption in the receiving room, m^2 (ft^2)

The sound absorption, A, in the receiving room is determined by measuring the reverberation time and using Sabine's equation (see Section 4.4.2.1). Care is taken to ensure that negligible impact sound is transmitted through paths other than the floor under test.

5.5.1.2 Field measurement of impact sound transmission

This is similar to the laboratory method, except that it is not possible to standardize the size and shape of the floor to be tested. The sound pressure level in the receiving room, resulting from the operation of the standard tapping machine on the floor above, is measured and it is normalised, either with respect to the absorption in the receiving room, in a similar manner to the laboratory method, or with respect to a standardized reverberation time of 0.5 seconds, which is taken to be typical for domestic sized, furnished rooms. Thus

$$L'_n = L_i \times 10 \log (A_0/A) \qquad [5.31]$$

where

L'_n = normalised impact sound pressure level, dB re 20 µPa

L_i = the average sound pressure level in the receiving room, dB re 20 µPa

A_0 = the reference sound absorption = 10 m^2 (100 ft^2)

A = the measured sound absorption in the receiving room, m^2 (ft^2)

$$L'_{nT} = L_i - 10 \log (T/T_0) \qquad [5.32]$$

where

L'_{nT} = standardized impact sound pressure level, dB re 20 µPa

L_i = the average sound pressure level in the receiving room, dB re 20 µPa

T = the measured reverberation time in the receiving room, s

T_0 = the reference reverberation time = 0·5 s.

5.5.1.3 Laboratory measurement of attenuation of floor coverings

This is an alternative technique whereby the effectiveness of different types of floor covering is assessed in terms of the reduction in transmitted sound compared to a standard 120 mm (4.7 in) thick reinforced concrete floor slab with an area of at least 10 m^2 (100 ft^2).[5.23] The standard specifies the size and fixings of the coverings tested, and the improvement is given as

$$\Delta L = (L_n)_0 - L_n \qquad\qquad [5.33]$$

where

ΔL = the reduction of impact sound pressure level, dB
$(L_n)_0$ = the normalised impact sound pressure level in the
      receiving room in the absence of floor covering
L_n = the normalised impact sound pressure level in the receiving
     room with the floor covering in place.

5.5.1.4 Criticisms of the ISO method for measurement of impact sound transmission

The sound transmitted to the receiving room when the floor above is impacted by the standard Tapping Machine is quite different to that when the floor is excited by footfalls, furniture movements, etc. The Tapping Machine tends to produce much higher sound pressure levels and the spectrum differs from that of footsteps; the rate of impact is also much faster than that of normal impact sounds on floors. The reasons for this are partly historical—because when the machine was first standardized impulse response sound level meters were not available and it was necessary to have a rapid impact rate in order to produce a relatively constant sound pressure level which could be measured in the receiving room. The high sound pressure levels resulting from the Tapping Machine are useful in cases of highly attenuating floors, in order that sufficient signal is transmitted to be measured above the ambient noise in the receiving room. However, it has been found that there are differences in the subjective ranking of the impact isolation of floors if tapping machine results are compared with footsteps. Olynyk and Northwood concluded that this was not of practical importance as the mis-ranking was only significant for floors with poor isolation.[5.24] However, there are serious theoretical criticisms of the method because the force applied to a floor surface by an impacter is not independent of the relative impedances of the impactor and the floor and in addition, many floors and floor coverings behave in a nonlinear manner, i.e. they behave differently under the type of forces applied by the Tapping Machine than under typical impact sources in buildings. Schultz and others have proposed alternative test methods using more realistic 'footstep' machines[5.25, 5.26] and research concerning this problem is continuing.

However, as is frequently the case in building acoustics, there are a great number of variables in practice—for example, in a typical dwelling, impact sound will be transmitted not only through the floor but through the connecting walls, columns and other building elements. Great care must be exercised in supervision if the potential impact sound attenuating properties of a floor system, as estimated from laboratory measurements are to be achieved. It is very difficult, and frequently impossible to carry out remedial work if errors have been made in construction.

5.5.2 Assessment of Impact Sound Transmission

The one-third octave band spectra measured in the receiving room when the ISO Tapping Machine is impacting the floor above may be rated using a standard reference curve. Since it is the *transmitted sound* that is being rated, the lower the levels, the better the attenuation of the floor. The reference curve is approximately the reverse of the airborne sound transmission curve, being flat from 100 to 320 Hz, a slope of −3 dB per octave from 320 to 1000 Hz and a slope of −9 dB per octave from 1000 to 3200 Hz, (see Fig. 5.10). The reference curve is shifted in steps of 1 decibel towards the measured sound pressure level curve until the mean unfavourable deviation is more than 1 dB but less than 2 dB. The value of the reference curve at 500 Hz is the single value rating number for the floor. If at any frequency the deviation exceeds 8 dB this should be reported.

Alternatively, the improvement (or decrement) over a reference curve, with the value at 500 Hz of 60 dB, may be stated as the *impact sound protection margin*.

5.6 BUILDING SERVICES NOISE CONTROL

The noise emitted by building services must be carefully examined and controlled if acceptable levels are not to be exceeded in the various occupied spaces. The acoustical design of such services is a specialized field, but building designers should be aware of the major sources and their control. As in all aspects of noise control in buildings, careful detailing and strict supervision is required to avoid problems that may not be amenable to correction in the completed project.

5.6.1 Air Conditioning and Mechanical Ventilation Systems

Air conditioning or mechanical ventilation is used in many buildings to provide thermal comfort and the necessary air quality. To be effective,

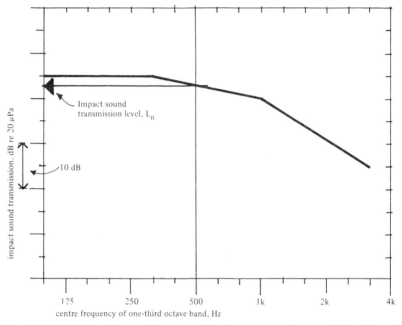

Fig. 5.10. Single number rating system for impact sound transmission through floors; when $L_n = 60$, the curve represents the reference values for impact sound transmission ISO 717/2.

air must be supplied in sufficient quantities to the different occupied areas of a building and the higher the velocity of the air, the smaller the size of the ducts needed for its transportation. In order to maximize the usable space in a building, whose overall dimensions are often limited by urban planning regulations, there is a tendency to use the smallest possible ducts, which requires the use of high velocity systems. (A low velocity system is classified as one in which the air speed is less than about 10 m/s (2 000 ft/min) and a high velocity system is one with higher speeds than this.) Since an elementary law of aerodynamic noise generation states that the emitted noise is proportional to the eighth power of the velocity of the air, high speed systems are potentially noisy. Fig. 5.11 illustrates some of the chief potential noise sources in ventilation systems.

In any ventilation system, a fan is used to move the air and this is the major potential source of noise. There are two main types of fan—the axial fan, which has most acoustic energy in the mid-frequencies, and

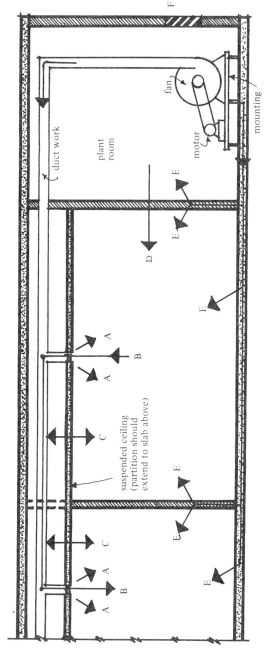

Fig. 5.11. Potential noise sources in ventilation systems: A) fan airflow and terminal device noise; B) cross-talk (both directions); C) duct break-out noise; D) airborne sound transmission; E) structure-borne sound re-radiated as airborne sound; F) intake and exhaust noise radiation to the community.

the centrifugal fan, which has a falling spectrum, with most energy in the low frequencies. Aerodynamic noise associated with fans has two sources: the first is rotation noise, caused by the interaction between the impeller blades and the surrounding air. An impulse is created each time a blade passes a certain point, this causes a tonal sound, the frequency of which may be found from the product of the fan speed, in revolutions per minute and the number of blades; thus, for a given number of blades, the slower the speed, the lower the tonal component. The second source of aerodynamic sound is vortex noise associated with turbulent flow arising from imperfect fan design. A fan produces least noise when it is operating at its maximum efficiency; thus it is essential that the correct fan is selected for the required duty—it is of no advantage to select an oversized fan.

The noise from the fan is transmitted along the ductwork, which usually includes a number of branches. It is important to realize that sound energy travels equally well with and against the direction of the air flow, so it is necessary to control noise emission from inlets as well as from outlets and exhausts. A general requirement for controlling ventilation noise is to design the ductwork to promote smooth air flow; abrupt changes of direction which result in turbulence should be avoided. The necessary air control devices, fire dampers, etc. should be carefully located. Turbulence is difficult to avoid in high velocity systems, which can also cause duct pulsation which is heard as a low frequency rumble which may be transmitted to lightweight suspended ceilings and partitions. It is usually necessary to line the ductwork with sound absorbent material, and, where duct runs are short, or where the acceptable sound level criteria are low, special attenuators may need to be used. These may take the form of *packaged silencers*, for which details of the attenuation provided should be available from the manufacture, or *plenum chambers*, which reduce noise transmission through impedance mismatch. The dimensions of packaged silencers and plenum chambers must be allowed for in the building design.

The air is finally delivered to each space through some type of terminal device, such as a grille, fan/coil terminal, induction coil unit and mixing box. Details of the noise generated by these devices, in terms of sound power levels under various conditions of air flow and pressure drop should be available from the manufacturers, but it is important to ensure smooth airflow approach conditions if these levels are not to be exceeded in practice.[5.27] Care is needed in the layout of ductwork so that *crosstalk* does not occur. This is the sound from one room that enters through an outlet, and through the ductwork in that room, and then travels along the duct into another room by-passing the attenuation provided by the dividing walls, floors, etc. A method of assessing noise reduction between two rooms through this flanking path is given by Craik and Mackenzie.[5.28] An additional path for sound transmission is referred to as duct 'break-out'—that is, the transmission of sound through the duct walls in or out of the rooms through which the duct passes. Since the transmission loss of typical duct walls is usually quite low, it is necessary to avoid duct layouts which only involve short lengths of duct between noisy and quiet rooms.

The location and design of the plant room is very important. Frequently the size allocated for the plant room is inadequate, allowing noise to bypass attenuators, e.g. if the supply and return air ducts are too close together. Fans, compressors, pumps, boilers, etc. should be provided with vibration isolating mountings, and it is also necessary to ensure that there are flexible connections between pipes, ducts and vibrating machinery. Whenever possible, the plant room should not be located near noise-sensitive spaces, and the radiation of sound from the plant room, the main air intakes and exhausts to the surrounding community should be controlled.

Although quantitative analysis of noise emission and control is possible for many elements of a mechanical ventilation system,[5.29] there are other aspects for which, in practical buildings, qualitative analysis only is available. It is therefore important that provision is made for acceptance testing and any necessary adjustments that need to be made before commissioning the system.

5.6.2 Plumbing and Drainage Systems

There are many potential noise sources in plumbing and drainage systems, such as fluid flow in pipes, valves and fittings, etc. Whereas the actual level of the noise may not be high, annoyance and even embarrassment may result, for example the noise of a toilet cistern filling heard in a nearby living room. If a liquid flows smoothly (viscous or laminar flow), it will be relatively quiet. However, if the flow is turbulent high noise levels may be emitted. Turbulent flow occurs when the Reynolds number is above about 2 200.[5.30] The Reynolds number is the product of the pipe diameter, the fluid flow velocity and density divided by the absolute viscosity of the liquid. For the pipe dimensions and the flow velocities typically used in domestic water supply systems, for example, the Reynolds number is over 50 000, and the flow is turbulent, and thus potentially noisy. The insertion of various valves and fittings may also contribute to turbulent flow.

Another source of noise in fluid flow is called cavitation. This occurs when a local restriction in the flow path results in a sudden increase in velocity with an accompanying reduction in pressure; at a particular velocity the pressure will be low enough for vapour bubbles to form. When the fluid passes the restriction the velocity decreases again and the pressure increases, leading to a sudden collapse of the vapour bubbles and extreme local pressure fluctuations. Cavitation may be minimized by correct design of valves and appliances.

Water hammer is another relatively common noise source; it occurs when the steady flow of a fluid is interrupted suddenly—for example when the solenoids of automatic washing machines operate. A sharp pressure rise occurs at the point of interruption and a steep shock wavefront may be reflected back and forth around the system many times. Automatic controls which may cause water hammer should be avoided; if they must be used it is necessary to provide some form of pressure absorbing device.

The small diameter pipes that are typically used in buildings are not efficient radiators of sound; it is therefore very important that they are not rigidly attached to efficient, monopole radiators such as walls and floors (see Section 1.6.3). If larger diameter pipes are used they may radiate a significant amount of sound energy. In this case the radiation may be reduced by applying laggings. An efficient method of reducing this type of noise is to surround the pipe with a porous, flexible material such as glass fibre or mineral wool and then to encase this with a thin impervious sheet, such as leaded vinyl. If possible, supply and waste pipes should be located within ducts, not embedded within the structure and they should be provided with resilient supports. Flexible connections between pumps and other machines should be provided also, to avoid transmission of structure-borne noise into the system.

Unfortunately, there is little quantitative information available regarding noise in plumbing and drainage systems. However, if good hydraulic practices are followed, which will promote non-turbulent flow, noise emission will be minimized. Heyman provided a good review of the principles of noise control in fluid piping systems,[5.31] and Gillam has provided a list of publications concerning service systems noise in buildings.[5.32]

5.6.3 Vertical Transportation Systems

Escalators and elevators must also be considered as potential noise sources. They should be provided with vibration isolating mountings, and elevators would have well-fitting doors to prevent sound transmission via the lift wells. (Fortunately, fire regulations now preclude the use of open wells and cages.) The elevator plant room must also be surrounded by highly attenuating construction, because of the noise associated with the electrical control gear, as well as the motors, etc. Careful planning to avoid juxtaposition of noise-sensitive areas and vertical transportation systems is required (Fig. 6.5).

5.7 SUMMARY

The theoretical background to the transmission of airborne and impact sound and vibration in buildings has been briefly discussed in this chapter. It has been shown that the behaviour of most building elements is strongly influenced by the frequency of the incident sound, and also, in many cases, the acoustic performance of a building element as measured in a laboratory may be severely modified when it is used in combination with other elements in a practical building. Thus, although the standard methods of measurement and assessment of these acoustic properties have been described, it is always prudent to include a factor of safety when selecting materials and construction systems if acoustical criteria are critical. More specific concerns relating to different building types are dealt with in Chapter 6.

REFERENCES

5.1. Beranek, L.L. The transmission and radiation of acoustic waves by solid structures, Chap.13. *Noise Reduction*, McGraw Hill, NY, 1960.
5.2. Quirt, J.D. Sound transmisison through windows I: Single and double glazing. *J.Acoust.Soc.Amer.* 72,1982, pp 834-844.
5.3. Quirt, J.D. Sound transmission through windows II: Double and triple glazing. *J.Acoust.Soc.Amer.* 74,1983, pp 534-542.
5.4. Ford, R.D. & Lord, P. Practical problems of partition design. *J.Acoust. Soc.Amer.* 43, 1968, pp 1062-1068.
5.5. Crocker, M.J. & Price, A.J. Sound transmission using statistical energy analysis. *J.Sound Vib.* 9,1969, pp 469-486.
5.6. Ver, I.L.& Holmer, C.I.,Interaction of sound waves and solid structures. Ch.11, *Noise & Vibration Control*, L.L.Beranek, ed. McGraw Hill, NY, 1971.
5.7. Brekke, A. Calculation methods for the transmission loss of single, double and triple partitions. *App.Acoust.* 14, 1981, pp 225-240.
5.8. Ellmallawany, A. Criticism of statistical energy analysis for the calculation

of sound insulation—Part 1, Single partitions, *App.Acoust.* 11, 1978, pp 305-312.

5.9. Ellmallawany, A. Criticism of statistical energy analysis for the calculation of sound insulation—Part 2, Double partitions. *App.Acoust.* 13, 1980, pp 33-41.

5.10. Ellmallawany, A. Improvement of the method of statistical energy analysis for calculation of sound insulation at low frequencies. *App.Acoust.* 15, 1982, pp 341-345.

5.11. Ellmallawany, A. Calculation of sound insulation of ribbed panels using Statistical Energy Analysis. *App.Acoust*, 18, 1985, pp 271-281.

5.12. Mullholland, K.A. & Lyon, R.H. Sound insulation at low frequencies. *J.Acoust.Soc.Amer.*, 54, 1973, pp 867-878.

5.13. Craik, R.J.M. Damping of building structures, *App.Acoust.* 14 1981, pp 347-359.

5.14. Tukker, J.C. Application of a measuring method for the dynamical behaviour of building structures. *App.Acoust.* 5, 1972, pp 245-264.

5.15. Brown & Maryon, Perception of wind movements. *Build.Mats. & Equip.* Aug. 1975.

5.16. Muster, D. & Plunkett, R. Isolation of Vibrations. Ch.13, *Noise & Vibration Control*, L.L.Beranek, ed. McGraw Hill, NY, 1971.

5.17. Bramer, T.P.C., Coles, G.J., Cowell, J.R., Fry, A.T., Grundy, N.A., Smith, T.J.B.,Webb, J.D. & Winterbottom, D.R. *Basic Vibration Control*, Sound Res.Labs. Suffolk, 1977.

5.18.—*Acoustics—Measurement of sound insulation in buildings and of building elements—Part III Laboratory measurements of airborne sound insulation of building elements: Part IV Field measurements of airborne sound insulation between rooms.* ISO 140-3 & 140-4. International Standards Organisation, 1978.

5.19. Lee, L.J. Development of a simplified field method of measuring sound insulation. *App.Acoust.* 18, 1985, pp 99-113.

5.20.—*Standard classification for determination of Sound Transmission Class.* ASTM E 413. Amer.Soc. for Testing and Materials.

5.21.—*Acoustics—Rating of sound insulation in buildings and of buidings elements—Part* 1: *Airborne sound insulation in buildings and of interior building elements.* ISO 717 Pt.1, 1982, International Standards Organisation.

5.22.—*Acoustics—Measurement of sound insulation in buildings and of building elements—Part IV: Laboratory measurements of impact sound insulation of floors* ISO 140-6 1978, International Standards Organisation.

5.23.—*Acoustics—Measurements of sound insulation in buildings and of building elements—Part VIII Laboratory measurements of the reduction of transmitted impact noise by floor coverings on a standard floor.* ISO 140-8, 1978, International Standards Organisation.

5.24. Olynyk, D. & Northwood, T. Subjective judgements of footstep-noise transmission through floors. *J.Acoust.Soc.Amer.* 38, 1965, pp 1035-1039.

5.25. Schultz, T.J. Alternative test method for evaluating impact noise. *J.Acoust.Soc.Amer.* 60, 1976, pp 645-655.

5.26. Broch, J.T. *Seminar on impact sound insulation test methods.* ELAB Report STF44 A82022 University of Trondheim, 1982.

5.27. Baade, P.K. Air conditioning systems for quiet buildings. *Internoise 77*, Zurich, 1977, p.B 393-400.

5.28. Craik, R.J.M. & Mackenzie, R.K. The transmission of sound by ventilation ducts—a design guide. *App.Acoust.* 14, 1981, pp 1-5.

5.29.—Chapter 35 Sound and Vibration Control. *ASHRAE Handbook*, Systems Volume. American Society of Heating, Refrigeration and Air Conditioning Engineers, New York. 1980.

5.30. Callaway, D.B. Noise in water and steam systems, Chapter 26, *Handbook of Noise Control*, C.W. Harris, ed. McGraw Hill, 1957.

5.31. Heyman, F.J. Acoustic performance tests and parameters for fluid piping system components; a critical evaluation of the state of the art, Part I, *App. Acoust.*4, 1971, pp 79-101, Part II *App.Acoust.*4, 1971, pp 155-173.

5.32. Gillam, A.J. Review of publications concerning services system noise in buildings. *App. Acoust.* 10, 1977, pp 33-48.

CHAPTER 6

Applications to Specific Building Types

6.1 INTRODUCTION

It must be emphasized that acoustical considerations form an important part of the design, detailing and construction of every building, in a similar manner to other technical aspects such as structures, building services, lighting, thermal comfort, etc. If acoustical mistakes are made (sometimes simply through lack of attention) it is often impractical to apply remedies in a completed building. Designers of larger building projects will usually seek expert advice from acoustical and other specialist consultants however, if they are to ensure the practical realization of their design concepts, it is essential that they have a thorough grasp of the principles involved. Builders, project managers and construction supervisors also need to understand the critical nature of acoustical detailing in order that the required effects are achieved.

In this chapter, acoustical problems and solutions relating to specific building types will be discussed. Since many problems are common to different building types there is a certain amount of repetition: however the advantages of completeness for each section appear to outweigh this disadvantage; cross-referencing to earlier chapters is also used extensively.

6.2 RESIDENTIAL BUILDINGS

6.2.1 Single-family Dwellings

In this type of building the main acoustical problem is usually the reduction of external noise intrusion, the source frequently being one of the various forms of transportation. (Wherever possible, sites adversely affected by road, railway and aircraft noise should be avoided.) Recommended sound levels inside dwellings are given in Section 3.2.2, and the corresponding criteria for outdoor sound levels depend on a) whether or not the rooms will be naturally ventilated and b) whether or not external areas associated with the dwelling, such as gardens, balconies, verandahs, etc. should also be protected from excessive noise intrusion. If the building is to be ventilated by windows or other unprotected openings, the reduction in sound level between outside and inside is small—it may be assumed that if 10% of the external envelope of a room consists of open windows, open doors, etc. then the maximum attenuation that can be expected is about 10 dB(A). The type of construction chosen for the remainder of the walls, the floor or the roof/ceiling system then becomes immaterial in such situations. If indoor criteria are to be satisfied for naturally ventilated dwellings, the external noise levels should not exceed 40 to 50 dB(A) daytime, or 35 to 45 dB(A) at night.

If the external noise sources already exist, for example if the site is near a developed road system, railway, or airport, measurements may be made in order to determine the external noise levels; Section 2.3 describes some typical community noise sources and their assessment and Section 2.4 points out some of the variables that affect sound propagation out of doors, and which must be taken into account if useful measurement results are to be obtained. If the sources do not yet exist, or if they are expected to change over the building's expected life, then prediction methods should be used to determine the expected external noise levels. Section 3.3.2 gives an example of predicting road traffic noise levels; Section 3.3.3 discusses the effect of barriers and Section 3.4 gives guidelines for assessing compatible land-use near airports. Railway and industrial noise assessment are included in Sections 3.5.2 and 3.6.3 respectively.

Once the external noise climate has been established, the possibility of using site planning to reduce noise intrusion should be explored. For example, if the main source is road traffic noise the building itself may be used as a shield against noise reaching external relaxation areas. In some circumstances it may be preferable to locate the building closer to the road, so that more external land at the rear is shielded, rather than to attempt to reduce the level at the exposed facade by increasing its distance from the road. (Only 3 dB(A) attenuation is predicted for each doubling of distance from a line source such as road traffic). Any topographical features that will act as a natural barrier should also be

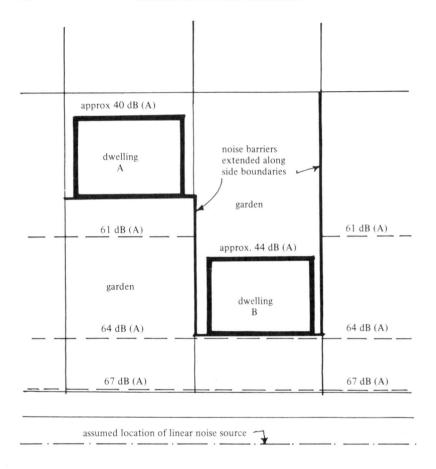

approx 40 dB (A)

dwelling
A

noise barriers
extended along
side boundaries

garden

61 dB (A)

61 dB (A)

approx. 44 dB (A)

garden

dwelling
B

64 dB (A)

64 dB (A)

67 dB (A)

67 dB (A)

assumed location of linear noise source

Fig. 6.1. Domestic building siting to shield external relaxation areas.

exploited. Fig. 6.1 shows a method of protecting the rear garden of the dwelling on the right by placing the building closer to the noise source and providing acoustically opaque side fences high enough to act as effective barriers.

The next stage is to consider the planning of the dwelling itself. Naturally other considerations such as aspect, view, etc. are present, but wherever possible, the least noise sensitive rooms, such as bathrooms, laundries, kitchens, etc. should be placed between the noise source and the more sensitive rooms such as bedrooms, living rooms, studies, etc.

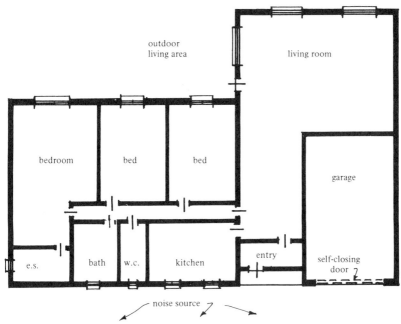

Fig. 6.2. Building planning to reduce external noise intrusion: suitable if natural ventilation required for cooling in warm climates.

There may be at least 20 dB(A) difference between the noise levels at front and rear facades of a dwelling exposed to road traffic noise. (See Fig. 6.2). However, if aircraft noise is the main external source, this method of shielding sensitive rooms is only possible in multi-level buildings, since the roof/ceiling construction will also be an important path for sound transmission.

Although the transmission of sound from one room to another in a single-family dwelling may be thought to be under the control of its occupants, when possible it is good practice to consider this aspect too. Plumbing noise can cause annoyance (and embarrassment), as can noise transmitted through ducted heating or air-conditioning systems. Impact noise on walls in kitchens and workrooms, and on lightweight floors may cause problems, and stereo, radio and TV listening may disturb other members of the household. Some of these problems may be overcome by sensible planning, however, others are related to construction and the location and fixing of building services.

The final consideration is the selection of the materials for the external and internal walls, the roof and the floors. Each country has its own forms of dwelling construction, and sound transmission loss data should be obtained for comparison purposes. (Appendix B gives some data for typical walls, windows, roofs, etc.). However, it is rare for the one material to be used for an external envelope, and a method of calculating the average attenuation of two or more different materials is given below. Since sound transmission loss data is provided as the reduction in decibels, which are logarithmic units, it is not possible to take a simple area-weighted arithmetic average, in decibels, to determine the overall sound transmission loss. It is necessary to determine the sound transmission coefficient for each component (for each frequency band of interest), and then to find the average, area-weighted sound transmission coefficient for the complete system; this is then converted to an average sound transmission loss in decibels. Although it is possible to make such calculations using a graph, a worked example is given below, in order to illustrate the significant effect of even a small area of a less attenuating material on the overall sound transmission loss.

Example

The external wall of a bedroom measures 2.8 m (9.2 ft) high x 5.0 m (16.4 ft) wide. It is to be built of 270 mm cavity brickwork. A 3 mm thick single-glazed openable window measuring 1.0 m (3.3 ft) high x 1.5 m (4.9 ft) wide is located in the wall. It is expected that the window will be half open for ventilation. The sound transmission loss data for the wall and the window is given in Table 6.1.

It is necessary to determine the average sound transmission loss at each frequency, using the following relationships:

$$\text{STL}_{av} = -10 \log t_{av} \qquad [6.1]$$

where

STL_{av} = the average sound transmission loss, dB
t_{av} = the average sound transmission coefficient
$\phantom{t_{av}}= (S_1 t_1 + S_2 t_2 + \ldots \ldots S_n t_n) / \Sigma S$
$S_{1, 2, n}$ = the surface areas of the different components
$t_{1, 2, n}$ = their respective sound transmission coefficients

At 100 Hz, firstly assuming that the window is closed:

$t_{wall} = 10^{-\text{STL}/10} = 10^{-39/10} = 1 \times 10^{-4}$

$t_{window} = 10^{-17/10} = 2 \times 10^{-2}$

TABLE 6.1
Sound Transmission Loss Data

	Centre frequency of 1/3 octave band, Hz															
	100	*125*	*160*	*200*	*250*	*320*	*400*	*500*	*630*	*800*	*1 k*	*1·25 k*	*1·6 k*	*2 k*	*2·5 k*	*3·2 k*
Wall	39	40	41	42	43	45	46	50	52	55	57	60	61	63	64	66
Window	17	15	16	15	17	14	18	19	23	22	22	21	18	18	17	20

TABLE 6.2
Calculated Overall Sound Transmission Loss, dB

Centre frequency of 1/3 octave band, Hz															
100	*125*	*160*	*200*	*250*	*320*	*400*	*500*	*630*	*800*	*1 k*	*1·25 k*	*1·6 k*	*2 k*	*2·5 k*	*3·2 k*
27	25	26	25	27	24	28	29	33	32	32	31	28	28	27	30

$$t_{av} = \frac{\{[(2 \cdot 8 \times 5 \cdot 0) - (1 \times 1 \cdot 5)] \times 1 \times 10^{-4}\} + \{(1 \times 1 \cdot 5) \times 2 \times 10^{-2}\}}{14 \cdot 0}$$

$$= [\underset{\text{(wall)}}{(1 \cdot 3 \times 10^{-3})} + \underset{\text{(window)}}{(3 \times 10^{-2})}]/14 \cdot 0$$

$$= 2 \cdot 2 \times 10^{-3}$$

$$STL_{av} = -10 \log (2 \cdot 2 \times 10^{-3})$$

$$= 26 \cdot 5 \text{ dB.}$$

Thus, although the window comprises only 10.7% of the total facade area, it has reduced the average sound transmission loss at this frequency from 39 dB to about 27 dB. Similar calculations are carried out for the other frequency bands and the results are shown in Table 6.2.

It can be seen that the transmission through the window determines the average performance of the whole facade, even when it is closed. If the window is open 50% to allow for natural ventilation, taking the transmission coefficient for an opening as 1.0, the average sound transmission loss at 100 Hz is as follows:

$$t_{av} = \frac{(12 \cdot 5 \times 1 \times 10^{-4}) + (0 \cdot 75 \times 2 \times 10^{-2}) + (0 \cdot 75 \times 1)}{14}$$

$$= (5 \cdot 48 \times 10^{-2})$$

$$STL = 12 \cdot 6 \text{ dB.}$$

This average sound transmission value will be found for the whole frequency range.

In order to put these calculations into context, it will be assumed that the traffic noise levels, L_{Aeq}, impinging on the facade are known. The A-weighted one-third octave band levels are shown in Table 6.3 (giving an overall level of 70 dB(A)).

The sound transmission loss values for the brick wall alone may be deducted from this traffic noise spectrum, in order to find the overall transmitted level (see Table 6.4).

These one-third octave band levels may then be combined, using the method shown in Tables 1.1, 1.2 & 1.3, giving the overall transmitted level of 19 dB(A). Thus the overall attenuation is $(70 - 19) = 51$ dB(A). (It should be noted that this is an approximate method of calculation, since no allowance has been made for facade reflection effects, or for the effect of the reverberation time of the receiving room.) For the wall

TABLE 6.3
Typical Road Traffic Noise Spectra A-Weighted One-Third Octave Band Levels, dB(A)

	Centre frequency of 1/3 octave band, Hz															
100	125	160	200	250	320	400	500	630	800	1 k	1.25 k	1.6 k	2 k	2.5 k	3.2 k
49	51	53	55	55	54	54	55	57	59	60	60	59	57	55	54

TABLE 6.4
Overall Transmitted Sound Levels

	Centre frequency of 1/3 octave band, Hz															
100	125	160	200	250	320	400	500	630	800	1 k	1.25 k	1.6 k	2 k	2.5 k	3.2 k
49	51	53	55	55	54	54	55	57	59	60	60	59	57	55	54
39	40	41	42	43	45	46	50	52	55	57	60	61	63	64	66
10	11	12	13	12	9	8	5	5	4	3	0	−2	−6	−9	−12

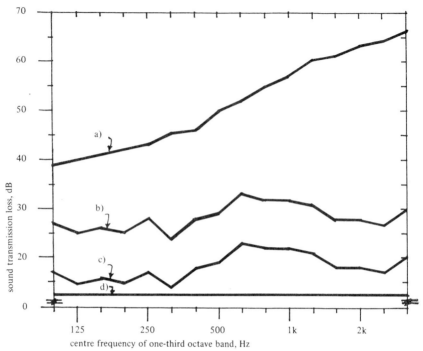

Fig. 6.3. The effect of open and closed windows on facade attenuation: a) 270 mm cavity brick wall; b) brick wall and window, window comprises 10.7% of total area; c) 3 mm single glazed window; d) as for b) but window 50% open (see text).

with the window, however, if the calculated average sound transmission loss values are deducted from the traffic noise spectrum the overall transmitted level is 41 dB(A) giving an overall attenuation of only (70–41) = 29 dB(A). In the open window situation the overall attenuation is only about 13 dB(A), giving a transmitted level of 57 dB(A). Fig. 6.3 illustrates this effect also.

In the above example it was assumed that there was only one facade exposed to the road traffic noise—this may not be a justified assumption, and it has been found that for single storied dwellings, or for the upper floors of two- or more storied buildings, the roof/ceiling may be an important transmission path.[6.1] In this case, the roof/ceiling should be treated as one of the components in the calculations.

The inclusion of the roof/ceiling in the calculation of envelope attenuation is naturally essential for rooms with ceilings exposed to aircraft noise. An alternative method of selecting suitable materials in order to achieve acceptable indoor sound levels is to assume that all components will admit the same amount of sound energy. This is the method given in Australian Standard AS 2021.[6.2] In Section 3.4.2 a method of estimating the aircraft noise level to be expected at a site near an airport is described. In the following example it is assumed that the maximum aircraft flyover level is 85 dB(A) and that a maximum indoor aircraft sound level of 50 dB(A) is acceptable (this is higher than that considered acceptable for road traffic or other continuous noise sources, because of the relatively short duration of the aircraft flyover); thus the required overall aircraft noise attenuation is ANR = (85–50) = 35

Example
The floor and the roof/ceiling of a bedroom measure 3 m (9.8 ft) x 4 m (13.1 ft). There are two external walls, each 2.8 m (9.2 ft) high, and in each wall there is a window measuring 1.0 m (3.3 ft) x 1.5 m (4.9 ft). The mid-frequency reverberation time in the room is 0.5 s. What aircraft noise attenuation is required for the walls, windows and roof ceiling if an equal amount of energy is to be transmitted by each component?
The following equation is used:

$$ANA_c = ANR + 10 \log [(S_c/S_f) \times (K/h) \times (2\,TN)] \qquad [6.2]$$

where

ANA_c = the required aircraft noise attenuation of the component
ANR = the overall aircraft noise attenuation required
S_c/S_f = the ratio of the area of the component to the floor area of the room
h = the height of the room, m (ft)
K = a constant = 3, h in m (= 9·8, h in ft)
T = the mid-frequency reverberation time in the room, s
N = the total number of components.

In this example, $N = 3$ (roof/ceiling, wall, window) and the Sc/Sf ratios of each component are given in Table 6.5.
In order to achieve an overall ANR of 35, the required ANAc for each component is as follows:

$$ANA_{c\,roof/ceiling} = 35 + 10 \log [1 \times (3/2{\cdot}8) \times 2 \times 0.5 \times 3]$$

TABLE 6.5
Component: Floor Area Ratios

Component:	S_0/S_f
Roof/ceiling	
$(3 \times 4)/(3 \times 4)$ [or $(9\cdot8 \times 13\cdot1)/(9\cdot8 \times 13\cdot1)$]	$1\cdot0$
Walls	
$[(3 \times 2\cdot8) + (4 \times 2\cdot8) - (2 \times 1 \times 1\cdot5)]/(3 \times 4)$	
$\{or\ [(9\cdot8 \times 9\cdot2) + (13\cdot1 \times 9\cdot2) - (2 \times 3\cdot3 \times 4\cdot9)]/(9\cdot8 \times 13\cdot1)\}$	$1\cdot38$
Windows	
$(2 \times 1 \times 1\cdot5)/(3 \times 4)$	
[or $(2 \times 3\cdot3 \times 4\cdot9)/(9\cdot8 \times 13\cdot1)$]	$0\cdot25$

$$(or\ 35 + 10 \log[1 \times (9\cdot8/9\cdot2) \times 2 \times 0\cdot5 \times 3]$$

$$= 40$$

$$ANA_{c\,walls} = 35 + 10 \log[1\cdot38 \times (3/2\cdot8) \times 2 \times 0\cdot5 \times 3]$$

$$(or\ 35 + 10 \log[1\cdot38 \times (9\cdot8/9\cdot2) \times 2 \times 0\cdot5 \times 3]$$

$$= 41$$

$$ANA_{c\,windows} = 35 + 10 \log[0\cdot25 \times (3/2\cdot8) \times 2 \times 0\cdot5 \times 3]$$

$$(or\ 35 + 10 \log[0\cdot25 \times (9\cdot8/9\cdot2) \times 2 \times 0\cdot5 \times 3]$$

$$= 34$$

Taking into account typical aircraft noise spectra, the required overall sound transmission loss is about 5 units higher, if expressed in terms of STC or R'_w. Thus materials having overall ratings of about STC or R'_w 45, 46 and 39 dB would be suitable.

6.2.2 Multi-family Dwellings

The general requirements for external noise attenuation as discussed in Section 6.2.1 above also apply to multi-family dwellings. In addition, care must be taken to avoid noise intrusion from other residents. The location of residents' and visitors' car parking areas, children's play areas, on-site recreational facilities, etc. must be carefully considered. In addition, if the building(s) are naturally ventilated, sound from one occupancy may travel to others through open windows if they are incorrectly located. Courtyard shaped buildings, although useful in excluding external noise, may lead to problems of this nature. Protruding wing walls, balconies, etc. are useful in reducing flanking transmission between external windows (Fig. 6.4)

In order to minimise the required attenuation between different occupancies it is advisable to plan the building so that similar rooms adjoin both horizontally and vertically; in other words, mirror-planning horizontally, and similar planning vertically, will ensure that bedrooms adjoin bedrooms, kitchens—kitchens, bathrooms—bathrooms, and living rooms—living rooms. Common hallways, stairs and lifts should not be located next to bedrooms (See Fig. 6.5). It is very important that plumbing and drainage pipes do not pass through living rooms and bedrooms unless located in highly sound-attenuating ductwork. Electrical conduits and outlets should not compromise the attenuation of common walls and floor/ceiling systems. Fortunately, these acoustical requirements are often satisfied through compliance with regulations for the prevention of spread of fire from one occupancy to another.

The walls and floor/ceilings between separate occupancies and between common areas and private ones should be chosen to have sufficient airborne sound attenuation. In many countries, the minimum attenuation is specified by regulation, usually as a single-number rating (see Section 5.4.2), and the system may be chosen from a list of constructions that are 'deemed-to-satisfy'. These constructions are those for which the Sound Transmission Loss has been measured in a laboratory (see Section 5.4.1), and which have been rated as complying with the relevant minimum requirements. However, it is evident that if the sound attenuation performance of a 'deemed-to-satisfy' form of construction (e.g. a masonry wall) is actually measured in a building it frequently fails to comply with the sound attenuation requirements. Langdon et al.[6.3] reported on measurements made by the British Building Research Establishment which showed that 55% of party walls and over 60% of party floors, built to comply with the regulations, failed to meet the minimum performance standards; it was also found that a majority of the residents found the sound insulation inadequate. There are several reasons for this poor sound attenuation performance in buildings: firstly, there may be flanking transmission, via other walls, floors, ceilings, etc., secondly, there may be poor workmanship, for example, inadequate mortar in brickwork or poor sealing of joints in dry-wall systems, etc.; in addition, the physical sizes of the wall or floor and the rooms may lead to resonant transmission that did not occur in the laboratory situation. The British

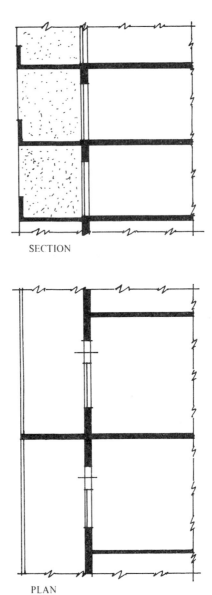

SECTION

PLAN

Fig. 6.4. Use of wing walls and balconies to reduce flanking transmission between external doors and windows.

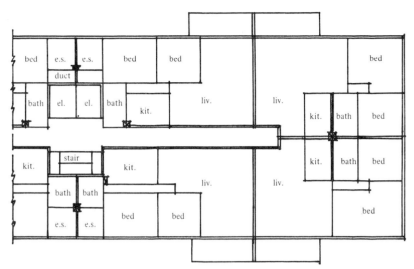

Fig. 6.5. Planning multi-family dwellings to reduce noise annoyance. Diagrammatic view of mirror-planning.

survey cited above found that many of the problems related to bad practice or to detailed design faults. Dry, lightweight forms of construction are particularly susceptible to inadequate sealing of joints. Whenever it is economically feasible, it is recommended that a 'factor-of-safety' is used, (as is the case in structural design) in order to allow for lower performance in practice. However, care must be taken that the potentially increased airborne sound reduction of the chosen construction is not negated by the same factors as outlined above; the whole system must be carefully examined, including the detailing of all joints, and critical parts of the construction should be supervised and the essential requirements for its acoustical integrity explained to the building supervisors and workmen.

Impact sound transmission should also be considered; in some countries there are requirements for minimum impact sound isolation of floors, usually described using a single number rating system (see Section 5.5.2). A floating floor system which incorporates a resilient layer is usually necessary to meet such requirements. Care is necessary to avoid bridging the resilient layer at walls and other junctions. Common stairways, halls, etc. should be separated from party walls and floors by resilient materials, to avoid transmission of sound into private spaces. Other common sources of impact sound are banging doors (which can be prevented by using door closers, resilient stoppers, etc.) and kitchen appliances, cooking, etc. It is recommended that party walls with benches attached should be of cavity construction.

6.3 COMMERCIAL BUILDINGS

6.3.1 Offices

The first consideration is the external noise climate at the building site. If the external noise sources already exist—for example, if the site is near a developed road system, railway, or airport—measurements may be made in order to determine the external noise levels; Section 2.3 describes some typical community noise sources and their assessment and Section 2.4 points out some of the variables that affect sound propagation out of doors, and which must be taken into account if useful measurement results are to be obtained. If the sources do not yet exist, or if they are expected to change over the building's expected life, then prediction methods should be used to determine the expected external noise levels. Section 3.3.2 gives an example of predicting road traffic noise levels; Section 3.3.3 discusses the effect of barriers and Section 3.4 gives guidelines for assessing compatible land-use near airports. Railway and industrial noise assessment are included in Sections 3.5.2 and 3.6.3 respectively.

The development of recommended sound levels inside offices is discussed in Section 3.2.1; they are primarily related to speech communication (for a description of the essential acoustical characteristics of speech see Section 4.2). Frequently, in an office situation, it is necessary to provide good speech intelligibility within a space and also good speech privacy between one space and adjoining spaces. This may be readily achieved if the spaces are physically contained, but if open-planning is used the acoustical problems are difficult to solve. Acceptable sound levels in offices should be regarded as minima as well as maxima, since lower ambient sound levels provide less masking noise, and thus less speech privacy. Table 6.6 gives some recommended ambient sound levels in offices and levels 5 dB(A) greater than these should be considered the maximum acceptable.[6.4]

The reverberation time in offices should not exceed about 0.4 seconds, except for larger spaces, such as conference rooms, where slightly longer reverberation is acceptable.

The actual, or expected external sound levels should be compared with the acceptable sound levels to determine the required attenuation

TABLE 6.6
Recommended Ambient Sound Levels for Offices

Type of Office	L_{Aeq}, $dB(A)$
General office areas, draughting areas, reception spaces, public spaces	40
Private offices, board and conference rooms	30
Computer rooms, typing pool areas, etc.	45

of the building envelope. If the building is to be ventilated by windows or other unprotected openings, the reduction in sound level between outside and inside is small—it may be assumed that if 10% of the external envelope of a room consists of open windows, open doors, etc. then the maximum attenuation that can be expected is about 10 dB(A). The type of construction chosen for the remainder or the walls, the floor or the roof/ceiling system then becomes immaterial in such situations. Office buildings are frequently provided with air-conditioning, in which case the necessary sound reduction between outside and inside can be achieved. The sealing of facade gaps, necessary for good thermal efficiency, is also an essential requirement for good airborne sound attenuation. However, a thermally efficient facade construction, e.g. factory sealed double-glazing with a very narrow space between the panes, is *not* acoustically efficient if there are important low frequency components in the external noise. The following example will illustrate this problem.

Example
A commercial building is to be located near a busy road, with about 2000 vehicles per hour; the sound level at the facade, L_{Aeq}, will be 75 dB(A). What is the estimated sound level inside the building if factory-sealed double glazing, consisting of two 6 mm (1/4 in) panes of glass 12.7 mm (1/2 in) apart, is used? Alternatively, what is the estimated sound level inside if the same thickness of glass is used in sealed frames with an air cavity of 76 mm (3 in)? The calculations are shown in Table 6.7.

Using the factory-sealed thermal glazing (line B) the overall transmitted sound level (line C) is 45 dB(A). If the wider spaced glazing is used (line D) the overall transmitted sound level is 38 dB(A), an

TABLE 6.7

Effect of Air Space on Attenuation of Road Traffic Noise by Glazing

Centre frequency of one-third octave band, Hz

	125	160	200	250	320	400	500	630	800	1 k	1·25 k	1·6 k	2 k	2·5 k	3·2 k	4 k
A	55	57	59	59	58	59	59	61	63	64	65	62	61	59	58	55
B	21	21	19	20	25	28	29	31	35	38	38	36	27	30	36	36
C	34	36	40	39	33	31	30	30	28	26	27	26	34	29	22	19
A	55	57	59	59	58	59	59	61	63	64	65	62	61	59	58	55
D	27	22	29	34	33	38	43	45	48	54	54	43	40	46	47	50
E	28	35	30	25	25	21	16	15	15	10	11	19	21	13	11	5

A = A-weighted traffic noise spectrum = 75 dB(A) overall.
B = STL for thermal double glazing.
C = transmitted sound level through B = 45 dB(A).
D = STL for 6:76:6 glazing.
E = transmitted sound level through D = 38 dB(A).

improvement of 7 dB(A). This latter level would also be suitable for general office areas, draughting rooms, etc., whereas the former level would only be suitable for computer rooms, typing pools, etc. Thus for the same surface density of glass, the wider air space between the panes gives superior traffic noise attenuation. However, there is a possible loss of floor space, owing to the thicker facade construction.

The required construction for internal partitions may be determined on the basis of providing speech privacy between adjoining offices. Speech privacy results if there is a sufficient noise to signal (speech) ratio. The source *speech level* depends on the distance over which communication takes place (the size of the source room), the vocal effort used (from quiet conversational speech to shouting) and also on the build-up of sound in the source room, which depends on its reverberation time (see Section 4.4.1). The *noise level* in the receiving room must not be too high, as discussed above, if speech communication within each office is to be satisfactory. The transmitted speech level depends on the sound attenuating characteristics of the partition, and also on its effectiveness as a sound radiator in the receiving room, i.e. if the receiving room is small and the common partition is relatively large, the average transmitted speech level will be higher, particularly if the receiving room is reverberant. A simplified method of selecting a suitable partition between two offices, based on speech privacy requirements is included in an Australian Standard;[6.5] this in turn is based on a method of Young.[6.6] It assumes that the background sound level is reasonably constant, which is true of backgrounds dominated by air-conditioning noise, but not necessarily true of background noise resulting from traffic.

Example
What airborne sound attenuation, expressed as its Sound Transmission Class, is required to provide confidential speech privacy between a private office, measuring 3 m (9.8 ft) x 3.5 m (11.5 ft) and a waiting room, measuring 3.5 m (11.5 ft) x 5 m (16.4 ft); the height of both rooms is 2.8 m (9.2 ft) and the background sound level in the private office is 30 dB(A) and in the waiting room is 45 dB(A). Both rooms are normally furnished, with carpets and curtains and an absorbent ceiling.
The basic equation used is as follows:

$$X = P + L_{A,V} + \Delta L_{A,S} - R_{eff} - N \qquad [6.3]$$

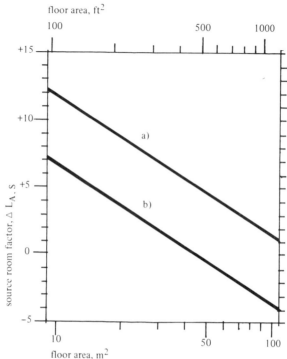

Fig. 6.6. Speech privacy calculations: speech level correction factor for size and reverberation time in source room: a) sparsely furnished, b) well-furnished.

where

X = the excess signal level (which should equal 0)

P = the noise excess over signal, or the privacy requirement (= 9 for normal privacy, 15 for confidential privacy)

$L_{A,V}$ = the average speech level at 1 m from the speaker in the source room (= 60 dB(A) for conversational speech, 66 dB(A) for a raised voice and 78 dB(A) for shouting)

$\Delta L_{A,S}$ = correction factor for size and reverberation time in the source room (see Fig. 6.6)

R = effective sound attenuation provided by the partition = (STC + K) or (R_w + K), where STC or R_w is the single number sound attenuation rating of the partition, and K is a correction factor for size of partition compared to floor area in the receiving room and the reverberation time in the receiving room (See Fig. 6.7)

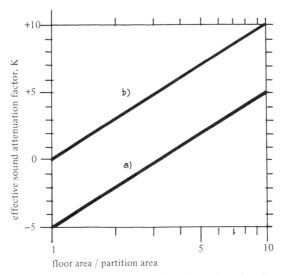

Fig. 6.7. Speech privacy calculations: correction factor for effective attenuation of partition between rooms: a) sparsely furnished receiving room, b) well-furnished receiving room.

N = background sound level in the receiving room
In this case, the value of STC (or R_w) is to be determined, thus the equation is rewritten:

$$R_{eff} = X + P + L_{A,V} + \Delta L_{A,S} - N$$

X will be taken as 0; P will be 15 (confidential privacy); $L_{A,V}$ will be taken as 66 dB(A) to allow for the raised voice situation. $\Delta L_{A,S}$ is found from Fig. 6.6 to be + 7 dB(A) and N in the receiving room is 45 dB(A):

$$R_{eff} = 0 + 15 + (66 + 7) - 45 = 43$$

$$R_{eff} = (STC + K)$$

and K is found from (area of receiving room floor/ area of common partition = 1.8) and Fig. 6.7 to be + 3, from which R_{eff} required = 40. Therefore a partition with an STC or an R_w rating of 40 would be suitable for this situation.

It is always advisable to make a check calculation in the opposite direction, to assess the speech privacy condition between the waiting room and the private office, in this case. The privacy condition will be taken as 9 for normal privacy and the speech level will be taken as 60

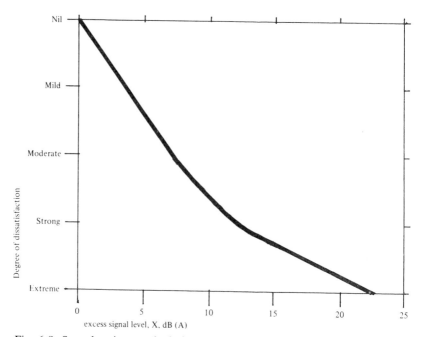

Fig. 6.8. Speech privacy calculations: relationship between excess signal level and speech privacy acceptability.

for normal conversational speech. $\Delta L_{A, S} = + 5$ from Fig. 6.6, $K = 0$ from Fig. 6.7 and $N = 30$ in the receiving room:

$$X = 9 + (60 + 5) - (40 + 0) - 30$$

$$= 4 \text{ dB(A)}.$$

From Fig. 6.8 an excess signal level of 4 dB(A) would lead to mild dissatisfaction with the speech privacy condition. This could be overcome by choosing a partition with a higher value for STC or R_w, or by increasing the background sound level in the private office by about 4 dB(A). The background sound level would then be 34 dB(A), which would still be acceptable. This illustrates the practical importance of using acceptable sound levels in offices as minima as well as maxima.

If the offices are not physically separated by partitions, the attenuation between one space and another occurs chiefly by distance; in such cases it is essential that all available surfaces are highly absorbent, otherwise reflected sounds will contribute to the signal level. For example it is predicted that at a distance of 4 m (13 ft) a speech signal level will be attenuated by about 10 dB(A) if the room has highly absorbing ceilings, thick carpets on the floor and 'acoustic' screens. If the room is reflective, the attenuation due to this distance would be only about 6 dB(A).[6.5] So-called acoustic screens have limited effectiveness in reducing sound transmission in open-planned spaces. The higher the screen, and the greater its lateral extent the greater will be the attenuation, but again, to be effective they should be surrounded by highly absorbent surfaces. Since open-planned offices usually have two dimensions very much greater than the third (the height) the normal assumptions regarding sound propagation as outlined in Section 4.4.1 do not apply. It is difficult to provide analytical solutions for such spaces, although experimental results have been published by several researchers.[6.7, 6.8] For example Pirn examined the effect of variables such as speaker voice level, orientation, background noise, speaker-to-listener distance, and barriers on the Articulation Index (see Section 4.2) He concluded that very small changes in any of these variables may change the situation from one of good speech privacy to one of good communication and vice versa.

Introduced masking noise has often been used in an attempt to improve speech privacy conditions in open-planned offices. However, it must be very carefully designed and maintained if it is not to be noticeable, and thus accepted by those working in the area. Warnock found that unless aural privacy is a prime consideration, no electronic masking system should be used.[6.9] Where loudspeakers are used to radiate masking noise they should be directed upwards in the plenum space to avoid marked changes in level as people walk around the room. There has been some controversy as to whether one spectrum of masking sound is better than another, but it is likely that provided that there are no prominent tonal components, any reasonably broad-band spectrum is satisfactory.

6.3.2 Shops and Retail Outlets

The acoustical requirements for these buildings have been considered less than those for offices. However, the acceptable sound levels indoors must again be determined by the requirements for speech communication. In turn, these depend on the type of retail establishment— in a self-service supermarket, for example, little communication is required and background music, advertisements, etc. are frequently the dominant sound sources. Ambient sound levels, L_{Aeq}, of 45 to 50 dB(A) are acceptable in such buildings. If detailed discussions are to take place, for example in specialty or technical stores, then the level should not exceed 40 dB(A).

Many retail stores are air-conditioned and thus suitable external envelope attenuation may be obtained (see Section 6.3.1 above). However, some smaller shops prefer to have to open doors so that there are no obstacles presented to customers. If they comprise strip-shopping on major roads it is very difficult to provide good communication conditions inside. The increased availability of automatically opening doors is one solution in these circumstances.

6.3.3 Hotels, Motels, etc

These buildings range in size from small, single storied motels with little else besides bedrooms and associated bathroom facilities to large multi-storied, air-conditioned hotels with restaurants, ballrooms, conference facilities, shopping concourses, etc. In the latter type of building careful planning is necessary to ensure that noise emitted in the public areas does not intrude into the guests' bedrooms.

The first consideration is the external noise climate at the building site. If the external noise sources already exist, for example if the site is near a developed road system, railway, or airport, measurements may be made in order to determine the external noise levels; Section 2.3 describes some typical community noise sources and their assessment and Section 2.4 points out some of the variables that affect sound propagation out of doors, and which must be taken into account if useful measurement results are to be obtained. If the sources do not yet exist, or if they are expected to change over the building's expected life, then prediction methods should be used to determine the expected external noise levels. Section 3.3.2 gives an example of predicting road traffic noise levels; Section 3.3.3 discusses the effect of barriers and Section 3.4 gives guidelines for assessing compatible land-use near airports. Railway and industrial noise assessment are included in Sections 3.5.2 and 3.6.3 respectively.

The recommended sound levels inside hotel and motel bedrooms are similar to those for private dwellings, i.e. 30 dB(A), L_{Aeq}, with 35 dB(A) the maximum acceptable. It is not recommended that lower sound levels than these are chosen, as this increases the possibility of loss of privacy between rooms. For recommended sound levels inside other parts of a hotel see the section dealing with the relevant building type (e.g. for conference rooms see Section 6.8.5, for offices see Section 6.3.1)

The actual, or expected external sound levels should be compared with the acceptable sound levels to determine the required attenuation of the building envelope. If the building is to be ventilated by windows or other unprotected openings, the reduction in sound level between outside and inside is small—it may be assumed that if 10% of the external envelope of a room consists of open windows, open doors, etc. then the maximum attenuation that can be expected is about 10 dB(A). The type of construction chosen for the remainder of the walls, the floor or the roof/ceiling system then becomes immaterial in such situations. Hotels and motels are frequently provided with air-conditioning, in which case the necessary sound reduction between outside and inside can be achieved. The sealing of facade gaps, necessary for good thermal efficiency, is also an essential requirement for good airborne sound attenuation. However, a thermally efficient facade construction, e.g. factory sealed double-glazing with a very narrow space between the panes, is not acoustically efficient if there are important low frequency components in the external noise. (See Section 6.3.1 for a worked example showing traffic noise transmission through thermal glazing compared to wide-spaced double glazing).

In order to determine the sound transmission loss required between bedrooms, the speech privacy condition should be calculated. A worked example of this calculation procedure is given in Section 6.3.1. Particular care should be taken to avoid annoyance in the guest areas arising from building services noise, especially from plumbing, elevators, etc. Section 5.6 describes some common sources of building services noise.

6.4 EDUCATIONAL BUILDINGS

6.4.1 Schools

The primary consideration in schools is the provision of good conditions for speech communication. In addition, there may be some rooms in which music is also important. Intrusion of sound from both external and internal sources must be considered. If possible, sites near major roads, railways or airports should be avoided. It is advisable to determine the external noise climate at the site at the relevant times of day. If the external noise sources already exist—for example, if the site is near a developed road system, railway, or airport—measurements may be made in order to determine the external noise levels; Section 2.3 describes some typical community noise sources and their assessment and Section 2.4 points out some of the variables that affect sound propagation out of doors, and which must be taken into account if useful measurement results are to be obtained. If the sources do not yet exist, or if they are expected to change over the building's expected life, then prediction methods should be used to determine the expected external noise levels. Section 3.3.2 gives an example of predicting road traffic noise levels; Section 3.3.3 discusses the effect of barriers and Section 3.4 gives guidelines for assessing compatible land-use near airports. Railway and industrial noise assessment are included in Sections 3.5.2 and 3.6.3 respectively.

Once the external noise climate has been established, the possibility of using site planning to reduce noise intrusion should be explored. For example, if the main source is road traffic noise the building itself may be used as a shield against noise reaching external areas requiring quiet, such as outdoor teaching areas. In some circumstances it may be preferable to locate the building closer to the road, so that more external land at the rear is shielded, rather than to attempt to reduce the level at the exposed facade by increasing its distance from the road. (Only 3 dB(A) attenuation is predicted for each doubling of distance from a line source such as road traffic). Any topographical features that will act as a natural barrier should also be exploited. Alternatively, playgrounds, sports fields, workshops, gymnasia (provided that the latter are not also used as assembly halls), etc. may be located between the noise source and the school classrooms. It is useful to list the noise sensitive and noise tolerant spaces and wherever possible to shield the former by the latter. However, some noise sensitive spaces are also noise sources on occasion, for example school auditoria, and this must also be considered when working on the initial planning of the building (See Fig. 6.9).

If the building is to be ventilated by windows or other unprotected openings, the reduction in sound level between outside and inside is small—it may be assumed that if 10% of the external envelope of a room consists of open windows, open doors, etc. then the maximum attenuation that can be expected is about 10 dB(A). The type of construction chosen for the remainder or the walls, the floor or the roof/ ceiling system then becomes immaterial in such situations.

Table 6.8 lists some recommended ambient sound levels for different parts of school buildings;[6.4] it is not advisable to design for lower levels than these, because of speech privacy considerations (see Section 6.3.1); levels 5 dB(A) higher than those tabled are the maximum desirable.

The reverberation time in most of these spaces should be about 0.4

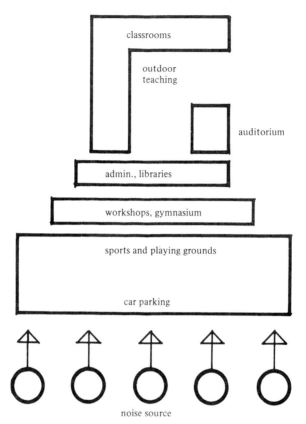

Fig. 6.9. Principles of siting of educational buildings to reduce external noise intrusion.

to 0.6 seconds. For the design of larger lecture rooms, assembly halls and theatres, see Sections 6.8.5, 6.8.6. and 6.8.2.

The required attenuation between classrooms and between classrooms and other spaces may be determined by comparing the acceptable sound levels with the expected source noise levels in the adjacent spaces. Where the source is primarily unaided speech, the speech privacy calculation method described in Section 6.3.1 may be used. Expected source speech levels (at 1 m distance) range from 60 dB(A) for conversational speech up to 78 dB(A) for shouting. It would be wise to assume a level of at least 66 dB(A) for a teacher's voice, and even higher levels in large spaces (these will be modified by the size and furnishings of the source room (see Fig. 6.6)). If speech is to be electronically amplified then the

TABLE 6.8
Recommended Ambient Sound Levels for Schools

Type of space	$L_{Aeq}dB(A)$
School classroom	35
Open-space teaching area	40
Audio-visual area	40
Teaching laboratories	35
Computer laboratories	40
Gymnasia	45
Lecture rooms, assembly halls and theatres, up to 250 seats	30
Assembly halls over 250 seats	25
Library reading areas	40
Offices	40

sound levels specified for the system should be used for the calculation. Live music has a wider dynamic range than speech (see Section 4.3) and maximum levels over 90 dB(A) may occur. Amplified music, particularly associated with audio-visual presentations, usually has a smaller dynamic range and also a narrower frequency band. It is important that good quality equipment with adequate high frequency response is used, otherwise the overall levels will be adjusted upwards in an attempt to provide good listening conditions, leading to unnecessarily high level low frequency components which are difficult to attenuate. It should be remembered that most single-number rating systems for airborne sound (STC, R_w) were developed in the context of speech and general household and office noise spectra (see Section 5.4.2)—they do not give a useful indication of the ability of the construction to provide low frequency sound attenuation. Wherever possible, the sound transmission loss characteristics of the proposed partition should be compared with the spectrum of the sound to be attenuated (see Section 4.3 for typical music spectra) and the overall transmitted sound level estimated.

Example
It is proposed to use a dry-wall partition between a music room and a school classroom. Assuming that there is no flanking transmission, and that the classroom measures 5 m (16.4 ft) wide by 7 m (23 ft) long and has a reverberation time of 0.6 seconds, what will be the average sound level in the classroom when the music room sound level is 89 dB(A) with a spectrum as shown below? The area of the partition is 4 m (13 ft) high × 5 m (16.4 ft) wide. The calculations are shown in Table 6.9

Adding the transmitted one-third octave band levels together gives 55 dB(A) overall. This should be corrected by 10 log S/A, where S is the area of the common partition and A is the absorption in the classroom. A is found from the reverberation time and Sabine's formula:

$$A = (KV)/T$$

where

A = is the total absorption in the room, m^2 (ft^2)
K = constant = $0 \cdot 161$, V in m^3 (= $0 \cdot 049$, V in ft^3)
V = volume of the room, m^3 (ft^3)

$A = (0 \cdot 161 \times 5 \times 7 \times 4)/0 \cdot 6$ {or $A = (0 \cdot 049 \times 16 \cdot 4 \times 23 \times 13)/0 \cdot 6$

$= 37 \cdot 6 \, m^2$ $= 400 \, ft^2$}

$10 \log S/A = 10 \log [(4 \times 5)/37 \cdot 6]$ {or $= 10 \log [(13 \times 16 \cdot 4)/400]$

$= -2.7$ $= -2 \cdot 7$}

Therefore the average sound pressure level in the classroom will be $(55 - 2.7) = 52$ dB(A). This is much higher than the recommended sound level of 35 dB(A), and a partition with better sound attenuation is required. The calculations will be repeated, using a double 'Camden' partition, as designed and used by the British Broadcasting Commission[6.10] (see Table 6.10).

Adding these transmitted one-third octave band sound levels together gives an overall level of 34 dB(A), which again should be corrected by 10 log S/A, giving $(34 - 2.7) = 31$ dB(A). This is quite satisfactory.

6.4.2 Tertiary Educational Establishments

These have the same basic requirements as schools, and Section 6.4.1 above should be consulted for details. There may be additional activities in tertiary institutions that should be considered, for example engineering laboratories may contain large machines typical of industrial premises and there may be medical and other laboratories with strict hygiene control requirements, typical of health buildings. In each case, the relevant section of this chapter should be consulted for detailed design guidelines.

There may also be a major auditorium associated with the institution; this should be designed in accordance with Sections 6.8.1, 6.8.2 or 6.8.6. It is most likely that any auditorium in this context will be multi-purpose

Acoustics and the Built Environment

TABLE 6.9
Transmission of Music Through Dry-Wall Partition

	Centre frequency of one-third octave band, Hz															
	125	*160*	*200*	*250*	*320*	*400*	*500*	*630*	*800*	*1 k*	*1·25*	*1·6*	*2 k*	*2·5*	*3·2*	*4 k*
A	65	66	68	69	70	75	78	79	80	81	81	80	79	76	73	71
B	20	22	25	28	30	33	35	36	38	40	40	39	38	40	45	50
C	45	44	43	41	40	42	43	43	42	41	41	41	41	36	28	21

A = assumed music spectrum, A-weighted.
B = attenuation of special dry-wall construction.
C = transmitted A-weighted sound levels.

TABLE 6.10
Transmission of Music Through Special Double Dry-Wall Construction

Centre frequency of one-third octave band, Hz

	125	160	200	250	320	400	500	630	800	1 k	1·25	1·6	2 k	2·5	3·2	4 k
A	65	66	68	69	70	75	78	79	80	81	81	80	79	76	73	71
B	37	38	43	49	49	49	56	60	65	67	71	74	78	76	78	77
C	28	28	25	20	21	26	22	19	15	14	10	6	1	0	−5	−6

in nature, and used for music, drama and even for ceremonial purposes. The latter may cause difficulties because of a requirement for part of the audience seating to be on a flat floor.

6.5 HEALTH BUILDINGS

Health buildings should be carefully sited away from major external noise sources if possible. Although community noise levels are usually lower at night than during daytime, and this may be taken into account for normal dwellings, it should be remembered that patients in hospitals may need to sleep during the day as well. It is advisable to determine the external noise climate at the site at the relevant times of day. If the external noise sources already exist—for example, if the site is near a developed road system, railway, or airport—measurements may be made in order to determine the external noise levels; Section 2.3 describes some typical community noise sources and their assessment and Section 2.4 points out some of the variables that affect sound propagation out of doors, and which must be taken into account if useful measurement results are to be obtained. If the sources do not yet exist, or if they are expected to change over the building's expected life, then prediction methods should be used to determine the expected external noise levels. Section 3.3.2 gives an example of predicting road traffic noise levels; Section 3.3.3 discusses the effect of barriers and Section 3.4 gives guidelines for assessing compatible land-use near airports. Railway and industrial noise assessment are included in Sections 3.5.2 and 3.6.3 respectively.

Once the external noise climate has been established, the possibility of using site planning to reduce noise intrusion should be explored. For example, if the main source is road traffic noise the building itself may be used as a shield against noise reaching external areas requiring quiet, such as patients' balconies, gardens, etc. Some parts of a hospital complex are not noise-sensitive, such as laundries, kitchens and general service areas, and these should be planned to protect wards, operating theatres, etc. from external noise sources.

If the building is to be ventilated by windows or other unprotected openings, the reduction in sound level between outside and inside is small—it may be assumed that if 10% of the external envelope of a room consists of open windows, open doors, etc. then the maximum

TABLE 6.11
Recommended Ambient Sound Levels for
Health Buildings

Type of space	L_{Aeq}, $dB(A)$
Single bed wards	30
Multi-bed wards	35
Operating theatres	30
Intensive care wards	40
Waiting areas, offices, etc.	40
Service areas	45

attenuation that can be expected is about 10 dB(A). The type of construction chosen for the remainder of the walls, the floor or the roof/ceiling system then becomes immaterial in such situations.

In cases where the building is to be air-conditioned, the attenuation provided by the external envelope may be estimated using the methods given in the two Examples in Section 6.2.1, and the transmitted sound levels compared with the recommended ambient sound levels given in Table 6.11.[6.4]

These levels may be increased by 5 dB(A) if necessary. The required attenuation between different parts of the complex may be determined by comparing the acceptable sound levels with the expected sound levels in adjacent spaces. Some assumptions must be made about activity noise levels in each type of space, or, if possible, measurements should be made of the noise emitted by similar equipment in existing premises. It would be wise to assume levels of at least 70 dB(A) in general ward areas (from conversation, etc.) and 80 dB(A) in service areas. Unless detailed information is available about source noise spectra, single-number rating data, such as STC or R_w, may be used to estimate the attenuation of airborne sound between spaces.

Impact sound transmission should also be considered; in some countries there are requirements for minimum impact sound isolation of floors for dwellings, usually described using a single number rating system (see Section 5.5.2). Similar impact sound isolation would be suitable for health buildings also. A floating floor system, which incorporates a resilient layer, is usually necessary to meet such requirements. Care is necessary to avoid bridging the resilient layer at walls and other junctions. Other common sources of impact sound are banging doors (which can be prevented by using door closers, resilient stoppers, etc.) and crockery and cutlery, etc.

There are usually many complex building services in large health buildings, and they should be required to meet the ambient sound level specifications appropriate to the difference spaces. See Section 5.6 for a discussion of building services noise.

6.6 PUBLIC BUILDINGS

These include transportation terminals (airports, railways, bus terminals, etc.), court houses, restaurants, museums, etc., and they are usually characterized by being visited by large numbers of members of the public.

6.6.1 Transportation Terminals

By their very nature these buildings must be located close to major transportation noise sources, and their design and construction need careful attention. Public areas, such as ticket sales, check-in, departure lounges, etc. should if possible have ambient sound levels not exceeding 45 dB(A), L_{Aeq}.[6.4] Levels of 5 to 10 dB(A) greater than these are permissible in areas where transactions take place, and if necessary, levels of 60 dB(A) can be accepted in waiting areas, departure lounges, etc. These recommended levels apply to relatively continuous sound such as emitted by road traffic noise. In the case of intermittent aircraft flyover noise, or train passby noise, these levels could probably be increased by a further 10 to 20 dB(A).

It is advisable to determine the external noise climate at the site. If the external noise sources already exist, for example, if the site is near a developed road system, or railway, measurements may be made in order to determine the external noise levels; Section 2.3 describes some typical community noise sources and their assessment and Section 2.4 points out some of the variables that affect sound propagation out of doors, and which must be taken into account if useful measurement results are to be obtained. If the sources do not yet exist, or if they are expected to change over the building's expected life, then prediction methods should be used to determine the expected external noise levels. Section 3.3.2 gives an example of predicting road traffic noise levels; and Section 3.3.3 discusses the effect of barriers. Railway noise assessment is included in Section 3.5.2. In the case of airport terminals, information should be sought regarding the type of aircraft that will be in use and their noise emission; maximum take-off and landing noise levels emitted by nearby large jet aircraft may exceed 100 or 110 dB(A).

Having determined the external noise levels, and the acceptable indoor sound levels, the required attenuation of the building envelope may be determined. If the building is to be ventilated by opening windows or if there are other unprotected openings such as large access doorways for transportation vehicles, the reduction in sound level between outside and inside is small—it may be assumed that if 10% of the external envelope of a room consists of open windows, open doors, etc. then the maximum attenuation that can be expected is about 10 dB(A). The type of construction chosen for the remainder of the walls, the floor or the roof/ceiling system then becomes immaterial in such situations.

If possible, passenger terminal buildings should be air-conditioned in order to attenuate the external noise levels sufficiently to achieve acceptable indoor sound levels. Methods of calculating the required attenuation are given in Section 6.2.1.

6.6.2 Court Houses

It is most important that court houses are designed to give good speech intelligibility in the court rooms themselves. Since many of the witnesses may have no experience of speaking in public the recommended background sound level is 25 dB(A), L_{Aeq}, with a maximum 5 dB(A) higher. It is also recommended that the mid-frequency reverberation time be limited to about 0.8 seconds, for rooms up to about 1000 m^3 (about 35 000ft^3) volume. (For details of designing for good room shape and correct reverberation time see Section 6.8.5).

Other parts of the building are more tolerant of noise, for example waiting areas, offices, etc. for which 40 dB(A) L_{Aeq} is acceptable. If possible the building should be planned so that the court rooms are shielded from external noise by less noise-sensitive areas.

The external noise climate at the building site should be determined, and, if possible, sites near major roads, railways or airports should be avoided. If the external noise sources already exist measurements may be made in order to determine the external noise levels: Section 2.3 describes some typical community noise sources and their assessment and Section 2.4 points out some of the variables that affect sound propagation out of doors, and which must be taken into account if useful measurement results are to be obtained. If the sources do not yet exist, or if they are expected to change over the building's expected life, then prediction methods should be used to determine the expected external noise levels. Section 3.3.2 gives an example of predicting road traffic noise levels; Section 3.3.3 discusses the effect of barriers and Section 3.4 gives guidelines for assessing compatible land-use near airports. Railway and industrial noise assessment are included in Sections 3.5.2 and 3.6.3 respectively.

The actual, or expected external sound levels should be compared with the acceptable sound levels to determine the required attenuation of the building envelope. If the building is to be ventilated by windows or other unprotected openings, the reduction in sound level between outside and inside is small—it may be assumed that if 10% of the external envelope of a room consists of open windows, open doors, etc. then the maximum attenuation that can be expected is about 10 dB(A). The type of construction chosen for the remainder of the walls, the floor or the roof/ceiling system then becomes immaterial in such situations. Court houses are frequently provided with air-conditioning, in which case the necessary sound reduction between outside and inside can be achieved. See Section 6.2.1 for an examples of calculating the attenuation of facades and building envelopes.

The required attenuation between court rooms, judges' chambers and legal interview rooms may be determined by using speech privacy calculations (see Section 6.3.1). Confidential speech privacy should be the goal for most of these spaces.

6.6.3 Libraries, Museums and Art Galleries

The public spaces in these buildings should be reasonably quiet, but they should not be too quiet otherwise they may inhibit visitors. An ambient level of 40 dB(A) L_{Aeq} is recommended. For recommended levels in offices and administration areas see Section 6.3.1. Guidelines for the design of any auditoria, seminar rooms, etc. associated with the complex are given in Section 6.8.5 . The reverberation times in the public spaces should not be too long, and should be related to the overall volume of each space. Since for practical purposes it is frequently necessary to use hard floor surfaces, and the spaces may be sparsely populated, care must be taken to provide acoustical absorbents on other surfaces. (See Section 4.6).

The external noise climate at the building site should be determined, and, if possible, sites near major roads, railways or airports should be avoided. If the external noise sources already exist, measurements may be made in order to determine the external noise levels; Section 2.3 describes some typical community noise sources and their assessment and Section 2.4 points out some of the variables that affect sound propagation out of doors, and which must be taken into account if useful measurement results are to be obtained. If the sources do not yet exist, or if they are expected to change over the building's expected life, then prediction methods should be used to determine the expected external noise levels. Section 3.3.2 gives an example of predicting road traffic noise levels; Section 3.3.3 discusses the effect of barriers and Section 3.4 gives guidelines for assessing compatible land-use near airports. Railway and industrial noise assessment are included in Sections 3.5.2 and 3.6.3 respectively.

The actual, or expected external sound levels should be compared with the acceptable sound levels to determine the required attenuation of the building envelope. If the building is to be ventilated by windows or other unprotected openings, the reduction in sound level between outside and inside is small—it may be assumed that if 10% of the external envelope of a room consists of open windows, open doors, etc. then the maximum attenuation that can be expected is about 10 dB(A). The type of construction chosen for the remainder of the walls, the floor or the roof/ceiling system then becomes immaterial in such situations. Libraries, museums and art galleries are frequently provided with air-conditioning, in which case the necessary sound reduction between outside and inside can be achieved. See Section 6.2.1 for an example of calculating the attenuation of facades and building envelopes.

6.7 INDUSTRIAL BUILDINGS

There are many different types and sizes of industrial building. The first requirement in the acoustical design of any industrial building is to ensure that employees do not receive noise exposures that may cause a permanent loss of hearing acuity. It is well known that excessive occupational noise exposure is related to progressive deterioration of high-frequency hearing, caused by irreversible damage to the hair cells in the cochlea (see Section 1.3). Individual sensitivity to such hearing loss is quite variable, so two people exposed to the same noise may have quite different physiological reactions. In countries where there are legislative controls for occupational noise it is usual to select an exposure level which will protect the majority of people over a working lifetime (of 40 years for example) from hearing loss sufficient to result in a handicap. The most noticeable effect of such handicap is a difficulty in understanding speech, particularly if there are other competing voices present, or if there is a high background noise level. A maximum level of 90 dB(A) L_{eq} for a maximum of 8 hours in each 24 hours has been incorporated in legislation in some countries. However, there is evidence to suggest that this is insufficiently protective, and for new buildings a maximum level of 85 dB(A) or less for an exposure duration of 8 hours is recommended. Exposure durations greater than or less than 8 hours are usually allowed for on an equal energy basis, so an exposure of 93 dB(A) for four hours is equivalent to an exposure of 90 dB(A) for eight hours (in the United States a 5 dB per halving of duration relationship is used, rather than the 3 dB implied in energy equivalence; e.g. 95 dB(A) for four hours is considered equivalent to 90 dB(A) for eight hours). Kryter published a general review of hearing loss, which should be referred to for more details.[6.11] Since many industrial noises are impulsive in nature, there has also been some research carried out into the best method of measuring and assessing the effect of this type of noise on hearing.[6.12] A comprehensive description of the methodology to be used in setting up a conservation program is given in International and other standards.[6.13, 6.14]

In order to determine the noise levels received by employees it is necessary to obtain data regarding the noise emission of the various machines and equipment that will be used, preferably in the form of one-third octave band sound power levels. The propagation of sound energy in a room is outlined in Section 4.4.1, and an example of how this may be applied in practice is given below:

Example

A source has a sound power level at 500 Hz of 103 dB re 10^{-12}W. The source is non-directional ($Q_\theta = 1$), the total room surface area is 400 m² and the average sound absorption coefficient at 500 Hz is 0.1. Would it be worthwhile increasing the total sound absorption in the room in order to reduce the noise level received by an operator located 3 m from the source? Using the relationship given in Equation 4.11:

$$L_p = 103 + 10 \log \left[(1/4\pi 3^2) + 4/(400 \times 0.1) \right]$$

$$= 103 + 10 \log \left[0.0088 + 0.1 \right]$$

$$= 93.4 \text{ dB re } 20 \,\mu\text{Pa}.$$

It can be seen that the reverberant sound field dominates at this position. Assume that the average sound absorption coefficient of the room surfaces is increased to 0.4:

$$L_p = 103 + 10 \log [(1/4\pi 3^2) + 4/(400 \times 0.4)]$$

$$= 103 + 10 \log [0.0088 + 0.025]$$

$$= 88.3 \text{ dB re } 20 \ \mu\text{Pa}$$

In this case, by increasing the average sound absorption coefficient at 500 Hz from 0.1 to 0.4 the sound pressure level at this frequency has been reduced by about 5 decibels. If this reduction were maintained over a sufficient frequency range to give an overall reduction of 5 dB(A) it may be worthwhile, particularly if a noise limit of 90 dB(A) must be complied with. However, it would probably be an expensive solution, since a large amount of material with very high sound absorbing properties would be required to achieve an average sound absorption coefficient of 0.4, as some of the surfaces, such as the floor, the windows, etc. would have very low coefficients, with much less than 0.1. In this case a reduction of 5 dB(A) in the noise emission of the source may be more economical, if technically feasible.

In some circumstances, in order to reduce excessive noise exposure, it may be more economical to isolate the people from the machines by providing sealed, ventilated control rooms, etc.

Any offices, laboratories, etc. associated with industrial buildings should be designed in accordance with the guidelines given in Sections 6.3.1 and 6.4.1. It is important to consider the attenuation between offices, amenities areas, etc. and machinery halls; frequently unacceptably high levels are found in such rooms because this has not been considered. The acceptable sound levels should be compared with the expected sound levels in the machinery hall and walls and floors selected with appropriate attenuating properties. If possible, the calculations should be made using one-third octave band data. Care must be taken that the potential attenuation is not reduced because of flanking transmission due to poor sealing, or penetration of piping or ductwork. Vibration should also be considered, preferably the source(s) should be provided with carefully designed vibration isolators (see Section 5.3.2). If this is not practicable, an alternative solution is to mount vibration sensitive rooms, such as control rooms, offices, etc. themselves on isolators; care must be taken not to bridge the isolation with rigid devices.

Whereas for most buildings the attenuation of the building envelope is selected with the goal of preventing excessive intrusion of noise from outside, in the case of industrial buildings it is sometimes necessary to ensure that excessive noise is not emitted to the surrounding community; see Section 3.6 for methods of assessing the impact of an industrial building on the surrounding community. If it appears that the noise emission will be excessive the best solution is to reduce the noise within the building, if possible. Alternatively, the building envelope construction should be designed to provide the necessary attenuation.

If the building is to be ventilated by windows, ventilation bridges or other unprotected openings, the reduction in sound level between outside and inside is small—it may be assumed that if 10% of the external envelope of a space consists of openings then the maximum attenuation that can be expected is about 10 dB(A). Open doors to loading docks, etc. must be included in the calculation of total open area. The type of construction chosen for the remainder of the walls, the floor or the roof/ceiling system then becomes immaterial in such situations.

In cases where the building is to be air-conditioned, the attenuation provided by the external envelope may be estimated using the methods given in the two Examples in Section 6.2.1.

6.8 AUDITORIA

This general term has been used to cover all spaces in which it is required to provide good conditions for listening and for producing wanted sound—whether speech or music. The general principles of room acoustics are discussed in Chapter 4 and some applications will be given in this section.

6.8.1 Concert Halls

These are the most acoustically difficult of all buildings, and there are still no guarantees of critical success; however, a considerable body of knowledge is now available for application in design. It may be assumed that an acoustical expert will be consulted during the design of any large concert hall; however, the building designer should also understand the principles involved.

Although it might seem to be obvious, the first most important factor to consider is the exclusion of external noise and noise from other parts of the building, including building services, from the concert hall itself.

The external noise climate at the building site should be determined, and, if possible, sites near major roads, railways or airports should be avoided. If the external noise sources already exist measurements may be made in order to determine the external noise levels; Section 2.3 describes some typical community noise sources and their assessment and Section 2.4 points out some of the variables that affect sound propagation out of doors, and which must be taken into account if useful measurement results are to be obtained. If the sources do not yet exist, or if they are expected to change over the building's expected life, then prediction methods should be used to determine the expected external noise levels. Section 3.3.2 gives an example of predicting road traffic noise levels; Section 3.3.3 discusses the effect of barriers and Section 3.4 gives guidelines for assessing compatible land-use near airports. Railway and industrial noise assessment are included in Sections 3.5.2 and 3.6.3 respectively.

Aircraft noise can be a particular problem with concert halls, since the large-span roofs required become very expensive if they must be massive to provide good airborne sound attenuation.

The ambient sound level inside a concert hall should not exceed 25 dB(A), and lower levels are preferable. If musical performances are to be recorded then it is also recommended that the low-frequency spectrum levels are controlled, to avoid overemphasis in subsequent replay. These ambient sound level criteria require very careful design of the ventilation system, and allowance must be made for low velocity air movement, which implies large-sized ductwork. The plant room must be placed well away from the concert hall itself, and it is important that it is sufficiently large to avoid cross-talk between intake and exhaust ducting. Plumbing and other services must be carefully designed and located. See Section 5.6 for a general discussion of services noise.

The attenuation required between the concert hall and other parts of the building is usually substantial; if there are several auditoria in the one complex source sound levels of over 100 dB(A) should be assumed for music rooms, and somewhat lower levels for drama theatres. Typically room-within-a-room construction is required to assure adequate airborne and structure-borne sound isolation between the concert hall and other auditoria (see Fig. 6.10 for an example).

The next stage is to determine the required audience capacity (if more than 2 000 people are to be accommodated it is very difficult to ensure good sound repetition over the whole audience area). The audience

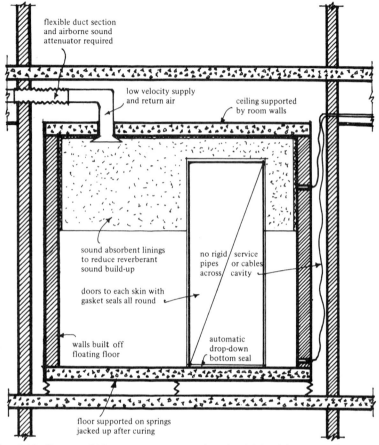

flexible duct section
and airborne sound
attenuator required

low velocity supply
and return air

ceiling supported
by room walls

sound absorbent linings
to reduce reverberant
sound build-up

no rigid / service
pipes / or cables
across / cavity

doors to each skin with
gasket seals all round

walls built off
floating floor

automatic
drop-down
bottom seal

floor supported on springs
jacked up after curing

Fig. 6.10. Room-within-a-room construction for high airborne and structure
borne sound attenuation.

seating area(s) should be laid out according to the guidelines for providing good sight lines (see Section 4.5). It is also necessary to take into
account local emergency exit requirements, which will affect the distance
between rows of seats, width of aisle and crossover lanes, etc. Although
audience proximity to the performers is desirable, the fan-shaped auditorium does not usually provide a sufficient number of reflected sounds
for musical enjoyment, and a more-or-less rectangular plan is recommended. In order to accommodate a large audience, without excessive distances between the rear rows of seats and the orchestral platform,

it is necessary to provide one or more balconies. These should also be designed to provide good sight lines to each seat. The average ceiling height will then determine the overall volume of the concert hall. Since the volume is an important parameter in determining the room's reverberation time, it should be selected in accordance with the latter's requirements. A mid-frequency reverberation time of about 1.8 seconds is recommended for a concert hall with a volume of 10 000m³ (350 000 ft³), rising to about 2.2 seconds for one with a volume of 50 000m³ (1 765 000 ft³) and dropping to 1.6 seconds for a hall with a volume of 5000 m³ (176 500 ft³). The audience itself is a strong sound absorber, particularly of high frequency components (see Section 4.6.6) and it is recommended that a volume per person of about 10 m³ (350 ft³) be provided if possible; in no case should the volume per person be below 7 m³ (250 ft³) per person. If 10 m³ (350 ft³) per person is assumed, a 10 000m³ (350 000 ft³) hall would accommodate 1000 people. In order to obtain a reverberation time of 1.8 seconds, the absorption must be limited to:

$$T_{60} = (KV)/A, \quad A = (KV)/T_{60}$$

where

T_{60} = reverberation time, s
K = constant, = 0.161, V in m³ (= 0.049, V in ft³)
V = volume in room m³ (ft³)

$A = (0.161 \times 10,000)/1.8$ [or $A = (0.049 \times 350,000)/1.8$
  $= 890$ m² $= 9,530$ ft²]

The mid-frequency absorption of an audience taken on a per capita basis is approximately 0.33 m² (3.5 ft²), thus an audience of 1000 people will itself provide approximately 330 m² (3 500 ft²) equivalent absorbent area, and all other surfaces should be limited to about 560 m² (6 030 ft²) at mid-frequencies. If the lower volume of 7 m³ (250 ft³) per person is considered, then the 10 000m³ (350 000 ft³) hall would accommodate about 1400 people, who would contribute approximately 460 m² (4 900 ft²) equivalent absorption, limiting the remaining surfaces to about 430 m² (4 600 ft²) at mid frequencies. It is recommended that the seating chosen for the audience has a similar sound-absorption characteristic to people, so that the variations in audience numbers will not markedly affect the reverberation time.

Although the audience contributes a major proportion of mid- and high-frequency component sound absorption, it is only about 50% as effective in the low frequencies. It is acceptable for concert halls to have slightly longer reverberation times at 125 Hz and below, but it is usually unnecessary to provide additional low-frequency absorption to avoid a 'boomy' condition. However, care must be taken that there is not too much low-frequency absorption resulting from extensive use of thin panels (see Section 4.6.3). A method of calculating reverberation time is given in Section 6.8.5. It should be remembered that it is a relatively easy task to reduce reverberation in an auditorium, by increasing absorption, but it can be very difficult to increase a reverberation time that is found to be too short, except by the use of electroacoustics, such as assisted resonance or ambiophony (see Section 6.8.10).

Reverberation time, as stated earlier, is a rather broad descriptor of the acoustical characteristics of an auditorium. It is important to ensure that each member of the audience receives good direct sound and several reflected sounds following within about 30 ms; the reverberant sound field should be diffuse and careful selection of room dimensions and room shape will help to ensure this. Geometrical analysis, carried out either manually or with the use of a computer should be used to determine the sequence of direct and reflected sound reaching the audience (see Section 4.5). Special diffusing surfaces may also be necessary. Schroeder has developed a useful method of designing wideband diffusers.[6.15] The performers' platform must also be carefully designed. The musicians need to hear their own instrument as well as those of their colleagues and they prefer to receive some sound reflected back from the audience area in order to perceive the 'room effect'. It is very important that the space in which the performers are located is part of the main space of the auditorium; placing musicians behind a proscenium arch on a stage is to be avoided, since much of the sound energy will be radiated into the wings and fly space, which are generally good absorbents of high frequency sound. If they must be placed on a conventional stage it is essential that a dense, rigid shell consisting of roof and wing walls is placed around the musicians. (See Fig. 6.11).

Because of the difficulty in making exact predictions about sound fields in concert halls, wherever possible it is useful to build physical acoustic scale models. These are typically built with a scale factor of 1/8 or 1/10 (although some have been built at a scale of 1/50). The wavelength of the test sounds have the same scale factor, which in practice means that the frequencies used must be eight or ten times higher than for full size. Thus if frequency range of say 100 to 5000 Hz is to be investigated,

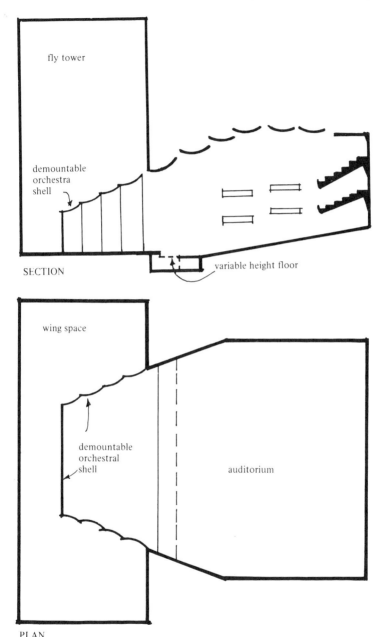

Fig. 6.11. Orchestral sound shell for use in proscenium-type theatres.

the test sounds will need to range from 800 to 40 000Hz or from 1000 to 50 000Hz. It is necessary to match the impedances of the surfaces (including the audience area) at the scale model frequencies to those of the real surfaces at full scale. One of the major problems to be overcome is that of absorption by air at the high model frequencies—this may be solved either by using very dry air, say about 3% relative humidity, or by filling the space with nitrogen rather than air. A general discussion of the principles involved in acoustical modelling was given by Brebeck et al.[6.16] and by Burd.[6.17] Provided that special transducers are used it is possible to 'play' orchestral music, speech, etc. in the model and to record it for subsequent replay and subjective assessment. In order to provide the necessary frequency transformation the sound played in the model is reproduced at eight or ten times the normal speed. A recent approach is to determine the impulse response of the model and to convolve this with the anechoic speech, music or other audio signals; it is claimed that this can give a good auditory impression of the effect of the room on the signal.[6.18]

Computer modelling is also increasingly being used to assist in the design of concert halls.[6.19] If possible, this should be carried out in three dimensions. It is relatively inexpensive, compared to physical models, and changes in shape and material characteristics may be readily carried out. However, it is not possible to 'listen' to the room's acoustics in this case.

Because of the difficulties in exactly predicting the acoustical characteristics of materials and spaces it is prudent to allow for several weeks 'tuning' after completion of the concert hall and before its official opening. (This may cause problems because of the usual time delays inherent in constructing large, complex buildings.) Measurements of reverberation time are easily made, and more sophisticated tests such as determination of the reflection sequence at the two 'ears' of dummy heads may also be required. (See Section 4.4.2.3). It is important that the absorption characteristics of an audience are allowed for if the tests are made in an unoccupied auditorium. Finally, test concerts, with real musicians performing and invited audiences are recommended, in order that any fine adjustments found necessary may be made. It is important for a concert hall's reputation that the local music critics approve of it— since every hall has its own unique sound, this is sometimes difficult to achieve, as the new hall will be different from the others that they have listened in.

6.8.2 Drama Theatres

The acoustical requirements for drama theatres are not quite as stringent as those for concert halls. However, the physical requirements for stages, lighting, projection, etc. may be in conflict with acoustical ones, and it is sometimes necessary to make compromises. The main criterion for a drama theatre is the promotion of good speech intelligibility. Since the power of the unaided human voice is limited (see Section 4.2), this requires a low ambient sound level and a limit to the size of the theatre to ensure that the voice will be sufficiently loud at all audience locations: this therefore places a limit on the number of people who can be accommodated. If electroacoustic sound reinforcement is to be used then larger audiences will receive adequate speech levels for good intelligibility. However, the distance at which facial expressions can be recognized is limited to about 20 m (65 ft), and people seated further away than this will not be able to perceive the drama adequately. It may be assumed that an acoustical expert will be consulted during the design of any major drama theatre; although this may not be the case for a small auditorium, it is always important that the building designer understands the main principles involved.

The external noise climate at the building site should be determined, and, if possible, sites near major roads, railways or airports should be avoided. If the external noise sources already exist measurements may be made in order to determine the external noise levels; Section 2.3 describes some typical community noise sources and their assessment and Section 2.4 points out some of the variables that affect sound propagation out of doors, and which must be taken into account if useful measurement results are to be obtained. If the sources do not yet exist, or if they are expected to change over the building's expected life, then the prediction methods should be used to determine the expected external noise levels. Section 3.3.2 gives an example of predicting road traffic noise levels; Section 3.3.3 discusses the effect of barriers and Section 3.4 gives guidelines for assessing compatible land-use near airports. Railway and industrial noise assessment are included in Sections 3.5.2 and 3.6.6 respectively.

Aircraft noise can be a particular problem with drama theatres since the large-span roofs required become very expensive if they must be massive to provide good airborne sound attenuation.

It is recommended that the ambient sound level inside a drama theatre should not exceed 25 dB(A), L_{eq}, and if possible, lower levels are preferable. This ambient sound level criterion requires very careful design of the ventilation system, and allowance must be made for low velocity air movement, which implies large-sized ductwork. The plant room must be placed well away from the drama theatre itself, and it is important that it is sufficiently large to avoid cross-talk between intake and exhaust ducting. Plumbing and other services must be carefully designed and located. See Section 5.6 for a general discussion of services noise.

The attenuation required between the drama theatre and other parts of the building is usually substantial; if there are several auditoria in the one complex source sound levels of over 100 dB(A) should be assumed for music rooms, and somewhat lower levels, say 85 to 90 dB(A), for drama theatres. Typically room-within-a-room construction is required to assure adequate airborne and structure-borne sound isolation between the drama theatre and other auditoria. See Fig 6.10 for an example.

The next stage is to determine the design of the performance area: if a conventional proscenium type stage is required it should be recognized that a considerable proportion of speech energy will be lost in the wings and fly tower; whenever possible, a large forestage should be provided and actors encouraged to use this part of the stage. An orchestra pit may also be required; typically this is designed to be flexible, in that additional audience seating may be placed over the pit if it is not required for musicians. However if this is done, care must be taken that these members of the audience have reasonable sight lines. The audience seating area(s) should be laid out according to the guidelines for providing good sight lines, taking into account the worst likely configuration of the proscenium opening (see Section 4.5 and Fig 4.9). It is also necessary to take into account local emergency exit requirements, which will affect the distance between rows of seats, width of aisle and crossover lane, etc. A fan-shaped auditorium has the advantage of seating large numbers of people fairly close to the stage, and, since for speech there is no special requirement for many reflected sounds to be perceived, this shape is quite suitable for drama theatres. However, the fan-angle should not be too wide (a maximum subtended angle at the speakers' positions of 140° is recommended)[6.20] as the high-frequency components of speech (which are important carriers of intelligibility) tend to be emitted in relatively narrow beams, and people sitting too much to one side of the actor have to rely on reflected sounds for these components. Since high frequencies tend to be absorbed by many materials, by the audience, and even by the air in the room, reflected sounds may well be deficient in the required components. For this reason, deep thrust stage and theatre-in-the-round configurations do not promote good speech intelligibility. One or more balconies may be necessary to accommodate the required audience within 20 m (65 ft) of the stage. Sight lines should be carefully designed so that each member of the audience has a clear view of the whole performance area.

The average ceiling height will then determine the overall volume of the drama theatre. Since the volume is an important parameter in determining the room's reverberation time, it should be selected in accordance with the latter's requirements. A mid-frequency reverberation time of about 0.8 seconds is recommended for drama theatres having a volume of about 1000 m³ (35 000 ft³) rising to about 1.2 seconds at 10 000 m³ (350 000 ft³). It is recommended that a volume of about 3 to 4 m³ (100 to 140 ft³) per person is allowed for drama theatres, any large volumes will require additional sound absorption to be used to keep the reverberation time low enough, and in addition, since the available sound energy will be distributed over a larger space, the average level will be lower.

Thus a theatre with an audience volume of 1000 m³ (35 000 ft³) will accommodate about 250 to 350 people, and one with a volume of 10 000m³ (350 000 ft³) would accommodate about 2 500 to 3 300 people. This sized audience would be much too large for unaided speech, and it would not be possible to accommodate them within 20 m (65 ft) of the stage.

In order to limit the mid-frequency reverberation time to 0.8 seconds in a 1000 m³ (35 000 ft³) drama theatre, there must be sufficient sound absorption, which may be calculated using Sabine's equation (see Section 4.4.2.1):

$$A = (KV)/T$$

where

A = the required sound absorption, m² (ft²)
K = constant = 0.161, V in m³ (= 0.049, V in ft³)
T = the required reverberation time, s

$A = (0.161 \times 1,000)/0.8$ [or $A = (0.049 \times 35,000)/0.8$
     $\cong 200$ m² $\cong 2,100$ ft²]

An audience of say 300 people will have a mid-frequency per capita absorption of approximately of 0.33 m^2 (3.5 ft^2) thus the additional absorption required will be:

A_{add} = 200 − (300 × 0.33)   [or A_{add} = 2,100 − (300 × 3.5)
   ≅ 100 m^2 ≅ 1,000 ft^2]

It is recommended that the seating chosen for the audience has similar sound-absorption characteristics to the people, so that the variations in audience numbers will not markedly affect the reverberation time. It is of course necessary to calculate the reverberation time for low, medium, and high frequencies, and a method is given in Section 6.8.5. The required amounts of absorbent material should normally be placed towards the rear of the room and particularly on surfaces that geometrical analysis has revealed may cause long-delayed reflected sounds to reach the audience (see Section 4.5).

The completed drama theatre should be subjected to acoustic testing before opening to the public. Reverberation times may be readily measured, as may the ambient sound levels (with the air-conditioning operating). In addition, speech intelligibility tests may be made, with listeners distributed over the audience area. One of the simple methods is to use the Modified Rhyme Test, in which a speaker reads out one of each group of six nonsense syllables, embedded in a short sentence, and the listener checks off the one perceived on a printed list. Each group of syllables has two common phonemes and one different one, e.g. ham, sam, cam, lam, mam, tam; and there are usually 50 of such groups in each test. Australian Standard AS 2822 describes the method to be used.[6.5]

Another method of assessing speech intelligibility in an auditorium is to use the Modulation Transfer Function. This measures the effect of the reverberation time and ambient noise on the original signal modulation. The method is described by Houtgast and Steeneken and a commercial instrument is available for the measurements.[6.21, 6.22] An advantage claimed for this method of assessment is a saving of time and personnel.

6.8.3 Opera Houses

The design of opera houses invites acoustical compromises—on the one hand it is necessary to provide the staging facilities required in drama theatres, on a large scale, and on the other hand it is necessary to provide good conditions for musical perception—the latter applying to the singers on stage and the orchestra as well as to the audience. It is well known that the traditional horseshoe shaped 19th Century opera house, as exemplified by the Teatro Alla Scala, Milano, the Staatsoper, Vienna and the Old Metropolitan Opera Theater in New York, cannot provide good sightlines to much of the audience (in fact there are considerable numbers of seats from which the stage is invisible). Lack of good sightlines means lack of direct sound, and thus intelligibility of speech and also of the sung text is very poor for these members of the audience. However, the distance between the front of the boxes as well as the front of the main floors is quite small, and conditions for sight and sound in these locations is very good. The audience seating tiers covers most of the walls, making them highly absorbent.

More egalitarian contemporary societies are less likely to accept an opera house with such variable quality in different parts of the audience area. An early attempt to provide a more even sound field was made by Wagner in the design of the Bayreuth Opera House, which was originally intended to have a single audience area (later compromised with a small balcony at the rear). The problem of blending voices on stage with the orchestra was solved by placing the orchestra pit underneath the stage, connected to the auditorium by a relatively narrow opening. This leads to a unique sound, as there is no direct orchestral sound received at all.

Contemporary audiences, familiar with home stereo recordings and with televised opera, appear to demand better intelligibility in opera singing, and frequently the libretto is sung in the local language. Because of the poor intelligibility in the Sydney Opera House, for example, the libretto is projected above the proscenium—this must be regarded as an acoustic failure! It may be assumed that an acoustical expert will be consulted during the design of any major opera house, however it is always important that the building designer understands the main principles involved.

The external noise climate at the building site should be determined, and, if possible, sites near major roads, railways or airports should be avoided. If the external noise sources already exist measurements may be made in order to determine the external noise levels; Section 2.3 describes some typical community noise sources and their assessment and Section 2.4 points out some of the variables that affect sound propagation out of doors, and which must be taken into account if useful measurement results are to be obtained. If the sources do not yet exist, or if they are expected to change over the building's expected life, then prediction methods should be used to determine the expected external noise levels. Section 3.3.2 gives an example of predicting road traffic noise levels; Section 3.3.3 discusses the effect of barriers and Section 3.4 gives guidelines for assessing compatible land-use near airports. Railway and industrial noise assessment are included in Sections 3.5.2 and 3.6.3 respectively.

Aircraft noise can be a particular problem with opera houses since the large-span roofs required become very expensive if they must be massive to provide good airborne sound attenuation.

It is recommended that the ambient sound level inside an opera house should not exceed 25 dB(A), L_{eq}, and if possible, lower levels are preferable. This ambient sound level criterion requires very careful design of the ventilation system, and allowance must be made for low velocity air movement, which implies large-sized ductwork. The plant room must be placed well away from the drama theatre itself, and it is important that it is sufficiently large to avoid cross-talk between intake and exhaust ducting. Plumbing and other services must be carefully designed and located. See Section 5.6 for a general discussion of services noise.

The attenuation required between an opera house auditorium and other parts of the building is usually substantial; if there are several auditoria in the one complex source sound levels of over 100 dB(A) should be assumed for opera houses and concert halls and somewhat lower levels, say 85 to 90 dB(A), for drama theatres. Typically room-within-a-room construction is required to assure adequate airborne and structure-borne sound isolation between an opera house auditorium and other auditoria. See Fig 6.10 for an example.

The geometrical design of the opera house should be basically similar to that for a drama theatre, in that a fully equipped stage area is required. (See Section 6.8.2). It is also necessary to provide for a large orchestra (of up to 100 players if large-scale operas are to be performed); in order to avoid a large distance between the edge of the stage and the first rows of audience seating, part of a large orchestra pit should extend under the stage itself. The audience seating area(s) should be laid out according to the guidelines for providing good sight lines, taking into account the worst likely configuration of the proscenium opening. (See Section 4.5). It is also necessary to take into account local emergency exit requirements, which will affect the distance between the rows of seat, width of aisle and crossover lanes, etc. If possible, all seats should be within 20 m (65 ft) of the stage—although it will usually be necessary to accommodate an audience of over 1 500 in an opera house for economic reasons and this will normally require the use of one or more balconies. Although a fan-shaped auditorium will assist in bringing the audience closer to the stage the fan-angle should not be too wide (a maximum subtended angle at the performers' locations of 140° is recommended)[6.20] and it must be remembered that incoherent (lateral) reflections which are desirable for musician perception are lost. A possible solution is to use a fan-shaped plan and to provide incoherent reflections from Schroeder diffusers[6.15] mounted on the upper part of the walls and the ceiling.

The reverberation time in an opera house should be longer than for a drama theatre, although not as long as for a concert hall of the same volume. This would indicate that a volume per person of the order of 5 m³ (175 ft³) would be suitable. This will determine the average ceiling height required. An opera house seating 1500 people would then require an audience volume of about 7 500 m³ (265 000 ft³), and the recommended mid-frequency reverberation time is 1.4 seconds for this volume. The audience itself would contribute approximately 0.33 m² (3.5 ft²) mid-frequency absorption per capita, or a total of 495 m² (5250 ft²) and the additional absorption required can be found using Sabine's equation (see Section 4.4.2.1):

$$A_{add} = [(0.161 \times 7{,}500)/1.4] - 495 \qquad [\text{or} [(0.049 \times 265{,}000)/1.4]$$
$$\cong 370 \text{ m}^2 \qquad\qquad\qquad - 5{,}250 \cong 4{,}000 \text{ ft}^2]$$

It is recommended that the seating chosen for the audience has similar sound-absorption characteristics to people, so that variations in audience numbers will not markedly affect the reverberation time. It is of course necessary to calculate the reverberation time for low, medium and high frequencies, and a method is given in Section 6.8.5. The required amounts of absorbent material should normally be placed towards the rear of the room, and particularly on surfaces that geometrical analysis has revealed may cause long-delayed reflected sounds to reach the audience (see Section 4.5).

Because of the difficulty in making exact predictions about sound fields in opera houses, wherever possible it is useful to build up physical acoustic scale models. These are typically built with a scale factor of 1/8 or 1/10 (although some have been built at a scale of 1/50). The wavelength of the test sounds should have the same scale factor, which in practice means that the frequencies used must be eight or ten times higher than for the full size. Thus if a frequency range of say 100 to 5000 Hz is to be investigated, the test sounds will need to range from 800 to 40 000Hz or from 1000 to 50 000Hz. It is necessary to match the impedances of the surfaces (including the audience area) at the scale model frequencies to those of the real surfaces at full scale.

One of the major problems to be overcome is that of absorption by air at the high model frequencies—this may be solved either by using very dry air, say about 3% relative humidity, or by filling the space with nitrogen rather that air. A general discussion of the principles involved in acoustical modelling was given by Brebeck et al.[6.16] and by Burd.[6.17] Provided that special transducers are used it is possible to 'play' operatic music, speech, etc. in the model and to record it for subsequent replay and subjective assessment. In order to provide the frequency transformation the sound played in the model is reproduced at eight or ten times the normal speed, recorded at this speed and then replayed at normal speed. A recent approach is to determine the impulse response of the model and to convolve this with anechoic speech, music or other audio signals; it is claimed that this can give a good auditory impression of the effect of the room on the signal.[6.18]

Computer modelling is also being used increasingly to assist the design of opera houses.[6.19] If possible, this should be carried out in three dimensions. It is relatively inexpensive, compared to physical acoustic models, and changes in shape and material characteristics may be readily carried out. However, it is not possible to 'listen' to the room's acoustics in this case.

Because of the difficulties in exactly predicting the acoustical characteristics of materials and spaces it is prudent to allow for several weeks 'tuning' after completion of the opera house and before its official opening. (This may cause problems because of the usual time delays inherent in constructing large complex buildings.) Measurements of reverberation time are easily made, and more sophisticated tests, such as determination of the reflection sequence at the two 'ears' of dummy heads may also be required. (See Section 4.4.2.3). It is important that the absorption characteristics of an audience are allowed for if the tests are made in an unoccupied auditorium. Finally, test operatic performances, with real opera singers and orchestral musicians performing and invited audiences, are recommended, in order that any fine adjustments found necessary may be made. It is important for the reputation of a new opera house that the local music critics approve of it—since every auditorium has its own unique sound, this is sometimes difficult to achieve, as the new opera house will be different to the others that they have listened in.

6.8.4 Churches, Cathedrals, Mosques and other Religious Buildings

The particular religious denomination for which the building is designed may govern the shape and size of the 'auditorium'. Western cathedrals have a long tradition of very large volumes, frequently with additional smaller volumes connected to them; they have very long reverberation times and traditional church music, with pipe organ accompaniment which exploits this type of acoustical characteristic.

However, in most religious buildings, speech intelligibility is also required, which is in direct conflict with the general lack of direct sound reaching the congregation and the predominance of the reverberant sound field. One method of overcoming this problem is to install an electroacoustic reinforcement system for speech, with carefully located loudspeakers which do not radiate sound into the high ceiling space. In some religious buildings, traditional Western church music does not have to be accommodated, in which case the reverberation time for a church with a volume of 1000 m³ (35 000 ft³) is 1.6 seconds, and for one with a volume of 50 000m³ (1.77 million ft³) it is 2.8 seconds; for speech reverberation time should be limited to 0.9 seconds and 1.4 seconds respectively (in the larger-volumed building speech reinforcement would be essential).

Wherever religious ceremonial requirements permit, a church or other auditorium should be designed according to the usual acoustical principles. That is, good direct sound from the source should reach each member of the congregation, who should not be located more than about 20 m (65 ft) away. The shape should be analysed using geometrical acoustics (see Section 4.5) and the ceiling height determined according to the volume necessary to achieve the required reverberation time. If speech is the main consideration, then the volume should be limited to about 3 to 4 m³ (100 to 140 ft³) per person; for traditional, reverberant church music, the volume should be at least 10 m³ (350 ft³) per person (this is similar to the recommendation for concert halls and even larger volumes may be preferable).

In many religious buildings the number of occupants is extremely variable; whereas the difference in total absorption may be overcome in theatres, etc. by using well-upholstered seating, this may conflict with tradition in a religious building. In order to prevent excessive reverberation during times of low occupancy it is recommended that alternative means of varying the absorption are used—for example unrolled absorbent banners, rotating panels that have absorptive material on one side and reflective material on the other, or even a sliding partition which may be used to reduce the volume of the room. A method of calculating the reverberation time in a room is given in Section 6.8.5. The required amounts of absorbent material should be placed on surfaces that geometrical analysis has revealed may cause long-delayed reflected sounds to reach the congregation.

6.8.5 Lecture Rooms, Conference Facilities

Although lecture rooms in schools and tertiary institutions are not very large, they frequently suffer from lack of acoustical consideration. The ambient sound levels are often too high, either because of noise intrusion from outside or from inadequately designed mechanical ventilation or air-conditioning systems. This also applies to conference facilities, particularly those provided as a part of a hotel complex. Sound reinforcement systems are often installed, but they also frequently suffer from poor design and poor quality components.

The external noise climate at the building site should be determined, and, if possible, sites near major roads, railways or airports should be avoided. If the external noise sources already exist measurements may be made in order to determine the external noise levels; Section 2.3 describes some typical community noise sources and their assessment and Section 2.4 points out some of the variables that affect sound propogation out of doors, and which must be taken into account if useful measurement results are to be obtained. If the sources do not yet exist, or if they are expected to change over the building's expected life, then prediction methods should be used to determine the expected external noise levels. Section 3.3.2 gives an example of predicting road traffic noise levels; Section 3.3.3 discusses the effect of barriers and Section 3.4 gives guidelines for assessing compatible land-use near airports. Railway and industrial noise assessment are included in Sections 3.5.2 and 3.6.3 respectively.

Ambient sound levels should not exceed 30 dB(A) L_{Aeq} in auditoria accommodating up to 250 people; for larger rooms that level should be 25 dB(A); in both cases, up to 5 dB(A) higher levels can be accepted if necessary. If the building is to be ventilated by windows or other unprotected openings, the reduction in sound level between outside and inside is small—it may be assumed that if 10% of the external envelope of a room consists of open windows, open doors, etc. then the maximum attenuation that can be expected is about 10 dB(A). The type of construction chosen for the remainder of the walls, the floor or the roof/ceiling system then becomes immaterial in such situations. Methods of calculating the attenuation provided by a building facade or a building envelope are given in Section 6.2.1.

The required attenuation between the lecture or conference room(s) and the rest of the building may be determined by comparing the acceptable sound levels with the expected sound levels in adjacent spaces. Some assumptions must be made about the activity noise levels in each space—it would be wise to assume at least 70 dB(A) from office type spaces and classrooms and up to 90 dB(A) if electronically amplified music is to be used. Unless detailed information is available about source noise spectra, single-number rating data, such as STC or R_w, may be used to estimate the attenuation of airborne sound between spaces.

Impact sound transmission should also be considered; in some countries there are requirements for minimum impact sound isolation of floors for dwellings, usually described using a single number rating system (see Section 5.5.2). Similar impact sound isolation would be suitable for lecture and conference rooms also. A floating floor system, which incorporates a resilient layer, is usually necessary to meet such requirements. Care is necessary to avoid bridging the resilient layer at walls and other junctions. Other common sources of impact sound are banging doors (which can be prevented by using door closers, resilient stoppers, etc).

Building services noise should also be considered, and they should meet the ambient sound level specifications: see Section 5.6 for a discussion of this type of noise and its propogation.

The shape of the room should be determined by laying out the audience seating area(s) according to the guidelines for providing good sight lines (see Section 4.5). In conference rooms it is also necessary to consider members of the audience as possible speakers, to be heard by their colleagues (frequently this is best achieved by using a roving microphone). Geometrical analysis should ensure that each member receives good direct sound and a few early reflections to increase loudness, if electronic reinforcement is not used. A fan-shaped auditorium has the advantage of seating large numbers of people fairly close to the podium, but the fan-angle should not be too wide (a maximum subtended angle at the speaker of 140° is recommended[6.20]) as the high frequency components of speech, which are important for intelligibility are radiated in relatively narrow beams from the speaker's mouth.

The recommended volume per person is about 3 to 4 m^3 (100 to 140 ft^3). This will determine average ceiling height. The mid-frequency reverberation time should be 0.7 seconds for a room with a volume of about 400 m^3 (14 000 ft^3) and 0.8 seconds for a room with volume of

TABLE 6.12
Sound Absorbent Coefficients of Materials and Audience

Material	Centre frequency of octave band, Hz					
	125	*250*	*500*	*1 k*	*2 k*	*4 k*
Brickwork	0·01	0·02	0·02	0·03	0·03	0·04
Plasterboard	0·20	0·15	0·10	0·08	0·04	0·02
Carpet floor	0·10	0·14	0·20	0·33	0·50	0·60
Doors	0·11	0·11	0·12	0·11	0·10	0·08
People (ea, m²)	0·19	0·28	0·33	0·35	0·37	0·39
People (ea, ft²)	2·0	3·0	3·5	3·8	4·0	4·2

about $1\,200\,m^3$ $(42\,000\,ft^3)$. The audience will provide approximately $0.33\,m^2$ $(3.5\,ft^2)$ mid-frequency absorption per person, thus the additional absorption required, for the larger room, seating 300 people may be calculated using Sabine's equation (see Section 4.4.2.1):

$$A_{add} = [(0.161 \times 1,200)/0.8] - 99$$
$$\cong 140\,m^2$$

$$[\text{or } A_{add} = [(0.049 \times 42,000)/0.8] - 1,050)$$
$$\cong 1,500\,ft^2]$$

It is recommended that the seating chosen for the audience has similar sound-absorption characteristics to people, so that variations in audience numbers will not markedly affect the reverberation time. It is of course necessary to calculate the reverberation time for low, medium and high frequencies and an example is given here. The required amounts of absorbent material should normally be placed towards the rear of the room, and particularly on surfaces that geometrical analysis has revealed may cause long-delayed reflected sounds to reach the audience (see Section 4.5).

Example
A small, air-conditioned lecture room is rectangular in shape with a flat floor and ceiling. It measures 5 m wide × 8 m long × 3.5 m high and accommodates 35 people. The floor is carpet on concrete, the ceiling is of plasterboard on battens, the walls are of painted brickwork and there are two doors measuring 2 m × 2 m. What is the predicted reverberation time with a full audience? Data for sound absorption coefficients for the

materials and audience is given in Table 6.12. The respective areas of the materials and the audience number are as follows:

Brickwork: $(2 \times 8 \times 3.5) + (2 \times 5 \times 3.5) - (2 \times 2 \times 2) = 83 \text{ m}^2$

[or $(2 \times 26.2 \times 11.5) + (2 \times 16.4 \times 11.5) - (2 \times 6.6 \times 6.6) = 893 \text{ ft}^2$]

Ceiling: (8×5) $= 40 \text{ m}^2$ [or (26.2×16.4) $= 430 \text{ ft}^2$]

Floor: $= 40 \text{ m}^2$ [or $= 430 \text{ ft}^2$]

Doors: $(2 \times 2 \times 2) = 8 \text{ m}^2$ [or $(2 \times 6.6 \times 6.6) = 87 \text{ ft}^2$]

People: 35

The calculations are shown in Table 6.13.

Extrapolating, for a volume of 140 m^3 ($4\,940 \text{ ft}^3$) the reverberation for speech should be about 0.6 seconds. It can be seen that although this is satisfied in the high frequencies, the reverberation is too long in the lower frequencies. The required additional absorption is as follows:

$$\Sigma A = (0.161/V)/T_{60}$$

$$\cong 37.6 \text{ m}^2$$

$$A_{add\ 125\ Hz} = 17.2 \text{ m}^2$$

$$A_{add\ 250\ Hz} = 13.6 \text{ m}^2$$

$$A_{add\ 500\ Hz} = 11.3 \text{ m}^2$$

$$A_{add\ 1\ kHz} = 5.5 \text{ m}^2$$

This requires a material with a good low frequency absorption: from Section 4.6 it will be seen that a panel (membrane) type absorber has this characteristic. Since the maximum absorption coefficient of such a material is usually about 0.5, about 34 m^2 would be desirable, with a resonant frequency around 125Hz. 10 mm plywood over an air space of about 40 mm would be suitable. The effect on other frequencies may be determined by recalculating the reverberation time. It will found that the absorption coefficient of this material is much less at higher frequencies, so the overall absorption should still be suitable. This panelling could be placed over some of the brick wall surfaces.

TABLE 6.13
Calculations of Reverberation Time

Material	Area S (m²)	Centre frequency of octave band, Hz											
		125		250		500		1 k		2 k		4 k	
		a	A	a	A	a	A	a	A	a	A	a	A
Brickwork	83	·01	0·8	·02	1·7	·02	1·7	·03	2·5	·03	2·5	·04	3·3
Ceiling	40	·20	8·0	·15	6·0	·10	4·0	·08	3·2	·04	1·6	·02	0·8
Floor	40	·10	4·0	·14	5·6	·20	8·0	·33	13·2	·50	20·0	·60	24·0
Doors	8	·11	0·9	·11	0·9	·12	1·0	·11	0·9	·10	0·8	·08	0·6
People	(35)	·19	6·7	·28	9·8	·33	11·6	·35	12·3	·37	13·0	·39	13·7
ΣA =			20·4		24·0		26·3		32·1		37·9		42·4
T_{60} = (0·161/V)/ΣA			1·1		0·94		0·86		0·70		0·59		0·53

6.8.6 Multi-purpose Auditoria

Multi-purpose auditoria, unless they are provided with sophisticated means of changing their configuration according to their use at a particular time, must involve acoustical compromises. In the worst case they may be used for classical music, popular music, drama, opera, cabaret, banquets, markets, etc. The flat main floor required for the last two or three activities is one of the chief acoustic problems, since it is not possible to provide the audience with good sight lines to the stage, and thus the direct sound will not reach them. Additionally, some type of stacking chair will be used, which does not usually provide as much absorption as a seated person, and thus the total absorption in the room will depend on the number of people present, unless other means of varying the absorption are provided.

It is advisable to decide on a clear list of priorities of use for the auditorium, and then to use the first priority as the chief determinant of room shape and volume. However, the other uses should also be considered: for example even if the main use is for drama, it would not be wise to use a complete fan-shaped plan, which is unsuitable for music. The volume should be large enough to support the longest reverberation time required, but provision should be made either to reduce this volume when the room is used for drama or to increase the total absorption in the room for this purpose. Alternatively, the auditorium could be designed for a shorter reverberation time to suit speech and a longer reverberation time simulated by electroacoustics, either using assisted resonance or ambiophony (see Section 6.8.10).

One difficulty that arises is the effect of a proscenium arch and stage house on musical performances. It is essential that an extended (retractable) forestage is provided so that at least the stringed instruments can be placed in front of the proscenium. A heavy orchestral shell should also be available to prevent orchestral sound being lost in the wings and fly space (see Fig. 6.11). It is a relatively simple matter to provide for a multi-positioned orchestra pit. It is more difficult to provide for an adjustably raked/flat main floor, but it is not impossible.

An extreme example of an auditorium which permits variation in size, shape and absorption is the IRCAM concert hall which is part of the George Pompidou Arts Centre in Paris.[6.23] The height of the ceiling may be varied between 11.5 m (37.7 ft) and 3.2 m (10.5 ft), the plan area may also be subdivided and the reverberation time can be altered between 1 s and 5.4 s at 1000 Hz. The walls and ceiling consist of three rotating prisms, which have absorbing, plane reflecting and diffusely reflecting

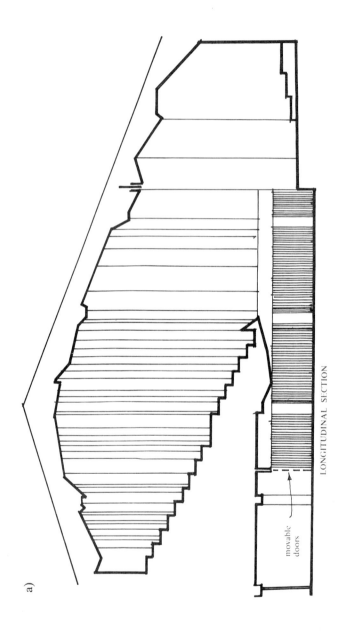

a)

movable
doors

LONGITUDINAL SECTION

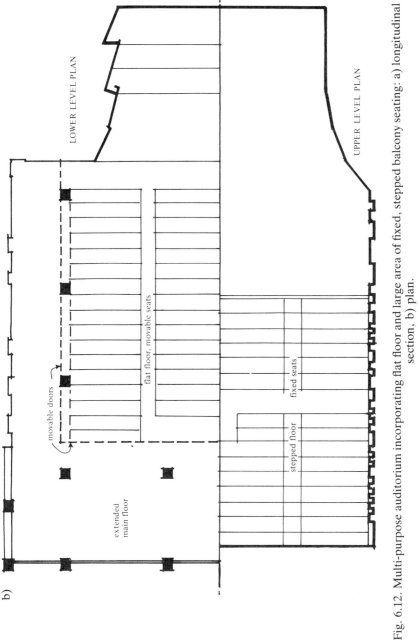

LOWER LEVEL PLAN

UPPER LEVEL PLAN

movable doors

flat floor, movable seats

fixed seats

stepped floor

extended main floor

b)

Fig. 6.12. Multi-purpose auditorium incorporating flat floor and large area of fixed, stepped balcony seating: a) longitudinal section, b) plan.

surfaces. Eventually the whole building will be controllable by computer so that particular configurations with known acoustical properties can be set up.

Another solution to the multipurpose, flat-floored auditorium, which does not require sophisticated alteration of shape or absorbents is to place much of the audience in fixed, stepped seating in balconies; these balconies may extend further than would normally be acceptable over the main floor area, provided that the rear part of the main floor is not used for concerts. An example of this is shown in Fig. 6.12 a) and b).[6.24]

For detailed design guidelines refer to Section 6.8.1 if the concert hall use is first priority, to Section 6.8.2 if the auditorium will mainly be used for drama and to Section 6.8.3 if it will mainly be used for operatic performances.

6.8.7 Stadia, Exhibition Halls, etc.

There have been several very large enclosed spaces, seating more than 10 000 people, built in recent years; they are designed to accommodate sporting events of various types, such as tennis, football, basketball, etc. However, they may also be used for concerts and theatrical performances. Because of the very large volumes of these spaces it is necessary to rely on sound reinforcement systems to radiate the sound to the audience—these need to be carefully designed by experts if they are to be successful.

Sightlines are usually satisfactory towards the central playing area, but unless some of the seating is closed off for theatrical performances some of the audience will be located behind the stage. The general guideline that no member of the audience should be more than 20 m (65 ft) from the performers cannot be complied with, and frequently people sitting in the further rows of the audience can make out very little of the acting on a stage.

Reverberation time needs to be controlled by the application of acoustical absorbents—usually the ceiling/roof is the only available surface.

6.8.8 Studios

The acoustical design of studios requires careful attention to detail.[6.25] It would be expected that for important studio complexes an acoustical expert will be consulted; however the building designer should have some understanding of the principles involved. There are several types of studio—the least critical are those used for television, as the visual

TABLE 6.14
Recommended Ambient Sound Levels in Recording Studios

	Centre frequency of octave band, Hz							
	32	*63*	*125*	*250*	*500*	*1 k*	*2 k*	*4 k*	*8 k*
Maximum sound pressure level, dB re 20 μPa	60	42	32	24	19	15	12	10	8

content tends to mask any aural deficiencies, and there are usually many people and pieces of equipment moving around inside them. The most critical studios are those used for audio recording.

The primary requirement is that the ambient sound level inside the studio is very low. In some countries the threshold of hearing is used as the criterion (see Section 1.3), although levels of about 20 dB(A) may be satisfactory for all but the most specialized applications. It is very important that low frequency components are controlled so that they do not become apparent after the recording–replay procedure. Table 6.14[6.4] gives recommended maximum background sound levels in recording studios.

The external noise climate at the building site should be determined, and, if possible, sites near major roads, railways or airports should be avoided. If the external noise sources already exist measurements may be made in order to determine the external noise levels; Section 2.3 describes some typical community noise sources and their assessment and Section 2.4 points out some of the variables that affect sound propagation out of doors, and which must be taken into account if useful measurement results are to be obtained. If the sources do not yet exist, or if they are expected to change over the building's expected life, then prediction methods should be used to determine the expected external noise levels. Section 3.3.2 gives an example of predicting road traffic noise levels; Section 3.3.3 discusses the effect of barriers and Section 3.4 gives guidelines for assessing compatible land-use near airports. Railway and industrial noise assessment are included in Sections 3.5.2. and 3.6.3 respectively.

It is also important to avoid the transmission of ground vibration, or vibration from other parts of a building into the structure of a studio. If this is likely to be a problem the vibration levels at the site should be measured and appropriate isolation provided.

The attenuation required between outside and inside and between the studio and other parts of the building can be determined by examining the spectra of the noise sources and comparing these with the recommended indoor sound level spectra (one-third octave band levels may be interpolated). In many cases it will be necessary to use special forms of construction, including room-within-a-room if the criteria are to be met (see Fig. 6.10). For information regarding a large number of practical constructions for broadcasting studios, see Randall et al.[6.10] Special doors with high sound attenuation are also required; they usually need special door furniture and good, long-wearing seals around jambs, head and bottom.

Air-conditioning is usually necessary and great care should be taken with the design of the system, including the routing of the ductwork, in order to avoid cross-talk occurring. Low velocity systems are preferred and allowance should be made for large-sized ducts and generous plant-room space, which should be located well away from the studios. There is also the need for a large number of cables, etc. to pass from studios to control rooms and it is essential that the openings are well-sealed around these.

It must be emphasized that all details of the construction must be very carefully worked out, and that strict supervision of all critical aspects must take place. A brick, or piece of plaster, or conduit bridging an air gap between the two skins of a cavity construction can seriously compromise the attenuation achieved—furthermore it is extremely difficult, if not impossible, to determine whereabouts in the construction the error lies.

The reverberation time in studios should be short, in order that recorded sound does not include unwanted reverberant components (if wanted, these are usually added electronically). A reverberation time of 0.2 seconds for small rooms, with volumes of about 50 m³ (1 700 ft³) is recommended, rising to about 0.7 seconds for a large film or television studio with a volume of 10 000m³ (350 000 ft³). These short reverberation times usually require the use of a considerable amount of sound absorbent materials. These must be carefully chosen in order to obtain a balanced reverberation time over the full frequency range of interest; they should also be placed evenly around the room.

A particular problem in small studios is the occurrence of widely spaced low-frequency room modes. These can lead to 'colouration'. See Section 4.4.2.2 for a method of calculating the frequencies at which these will occur. Helmholtz resonator type absorbers may be used, tuned to modal frequencies, to reduce their effect (see Section 4.6.4).

6.8.9 Open Air Auditoria

The ancient Greek theatres have been subjected to a considerable amount of acoustical analysis, one of the most important being the Theatre of Epidauros, which can seat a total of 14 000 people.[6.26] Cremer[6.27] used some of the principles of these theatres in his design of the Berlin Philharmonic Hall. The rake of the audience seating at Epidauros is a little over 26° from the horizontal, which gives good direct sound above grazing incidence. These theatres also had a reflective wall at the rear of the stage which provided early reflected sound. The focusing effects of circular, or near-circular seating are mitigated by the absorption provided by the audience itself. When originally constructed the ancient Greek and Roman theatres were in very quiet environments.

Contemporary outdoor theatres must be very carefully sited if external noise intrusion is not to be a problem—there is no external envelope to provide attenuation. The external noise climate at the building site should be determined, and it is essential that sites near major roads, railways or airports should be avoided. If the external noise sources already exist measurements may be made in order to determine the external noise levels; Section 2.3 describes some typical community noise sources and their assessment and Section 2.4 points out some of the variables that affect sound propagation out of doors, and which must be taken into account if useful measurement results are to be obtained. If the sources do not yet exist, or if they are expected to change over the building's expected life, then prediction methods should be used to determine the expected external noise levels. Section 3.3.2 gives an example of predicting road traffic noise levels; Section 3.3.3 discusses the effect of barriers and Section 3.4 gives guidelines for assessing compatible land-use near airports. Railway and industrial noise assessment are included in Sections 3.5.2 and 3.6.3 respectively.

A sound reflecting shell is usually provided over the platform area, and this should be designed to supplement the direct sound with short-delayed reflections to the audience area. Fan-shaped audience seating areas are frequently used in order to place as many people as possible close to the performers. However, the fan-angle should not be too wide. It is essential that the seating is raked to provide good sight lines, and thus good direct sound paths.

Usually, the natural sound is supplemented with electroacoustic reinforcement. If a large audience area is to be covered it is necessary to use a number of loudspeaker locations and the Haas effect must be taken into account and electronic time-delays incorporated in order that the sound will appear to originate from the natural source. (See Section 4.5).

6.8.10 Sound Reinforcement in Auditoria

This is a specialized field and experts should always be consulted. However, it is necessary for the building designer to understand that loud speakers must be correctly located if acoustical effects are to be realized. The simplest form of sound reinforcement is that used for lecturers, speakers, etc. and it comprises a microphone, an amplifier and one or more loudspeakers. There is a wide range of quality available in each of these components and it is essential that all are of similar quality, otherwise money will be wasted. The basic requirements to be satisfied are a) an adequate frequency range and b) adequate sound power to radiate sufficient sound level without distortion.

It is essential that conflict between the visual and audio directional location of the source does not occur. Use may be made of the Haas effect to avoid this problem (see Section 4.5). In its simplest form, the loudspeaker should be located further from the audience than the real source—however, if this is attempted by placing the loudspeaker behind the real source, a feed-back loop between the microphone and loudspeaker may be set up, leading to howling. In more sophisticated installations, electronic time delays are used to compensate for the fact that the loudspeaker is closer to the listener than the real source. (See Fig. 6.13).

Another use of electroacoustics in auditoria is to create a longer reverberation time. A system known as Assisted Resonance was developed by Parkin,[6.28] and used in the Royal Festival Hall, London, when it was found that the low-frequency reverberation was too short for music. The system has since been used in other auditoria, particularly in multi-purpose auditoria which have been designed primarily for speech, but which are also used for music. The system employs a large number of channels which each consist of a microphone located within a Helmholtz resonator and which responds to a narrow band of frequencies, placed at an antinode (a maximum particle vibration position) for that band of frequencies; the signal is amplified and re-radiated from a loudspeaker. By controlling the amplifier gain the reverberation time of each natural room resonance is increased, and thus is the overall reverberation time of the room. Although expensive to install, and needing expert control during use, it is claimed that it is cheaper than

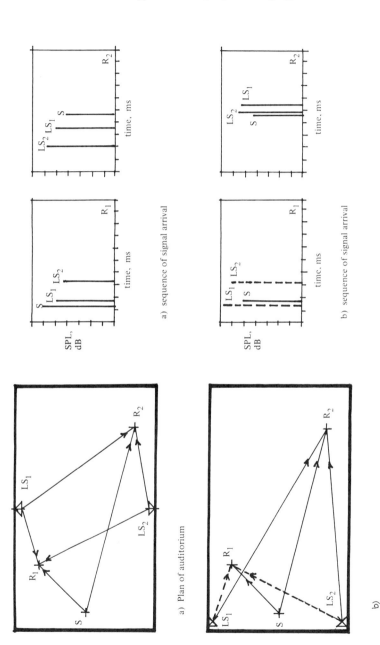

Fig. 6.13. Location of loudspeakers and apparent location of sound source: a) loudspeakers distributed around room; apparent location at S for R₁, but at LS₂ for R₂ unless appropriate electronic time delays incorporated; b) loudspeakers behind source, S; apparent location at LS₁ for R₁ unless LS₁ directed away from front rows of audience; apparent location at S for R₂.

building the additional volume necessary to achieve a longer reverberation suitable for classical music. It is also one of the few methods available to correct this problem in an existing building. Another method of extending reverberation in a room is to use ambiophony. In this case one or more microphones are located close to the sound source(s) and amplified, time-delayed signals are radiated from loudspeakers located in a large number of positions around the auditorium. This attempts to reproduce the effect of a large number of reflected sounds from the upper walls and ceiling.

6.9 SUMMARY

In this chapter an attempt has been made to apply the theories and guidelines of the earlier chapters to the acoustical design of a number of discrete building types. Certain portions of the text have been repeated in order that the building designer does not have to hunt through irrelevant building types in order to find the information required.

In all cases, it must be reiterated, the required acoustical performance will not be achieved unless the building designer, the contractor and the tradespeople involved understand why various details are required, and expert supervision is available to ensure faithful execution.

REFERENCES

6.1. Lawrence, Anita, Noise reduction of facades—more measurement results. 12*th Int. Congress on Acoust.* Toronto, 1986, E2-3.
6.2.—*Acoustics—Aircraft noise intrusion—building siting and construction*, AS 2021-1985 Standards Association of Australia, 1985.
6.3. Langdon, F.J., Buller, B. & Scholes, W.E. Noise from neighbours in multi-storey flats, *Building Research Establishment News*, Summer, BRE, U.K., 1983, p 13.
6.4.—*Acoustics—Recommended design sound levels and reverberation times for building interiors.* AS 2107-1987, Standards Association of Australia.
6.5.—*Acoustics—Methods of assessing and predicting speech intelligibility and speech privacy.* AS 2822-1985, Standards Association of Australia, 1985.
6.6. Young, R.W. Revision of the speech-privacy calculations. *J.Acoust.Soc. Amer.* 38, 1965, pp 524-530.
6.7. West, M. & Parkin, P. The effect of furniture and boundary conditions on the sound attenuation in a landscaped office; Part 1 *App.Acoust.* 8, 1975, pp 43-66; Part 2 *App.Acoust.* 11, 1978, pp 171-218.

6.8. Pirn, R. Acoustical variables in open planning. *J.Acoust.Soc.Amer.* 49, 1971, pp 1339-1345.

6.9. Warnock, A.C.C. Acoustical Privacy in the landscaped office. *J.Acoust. Soc.Amer.* 53, 1973, pp 1535–1543.

6.10. Randall, K.E., Meares. D.J. & Rose, K.A. *Sound insulation of partitions in Broadcasting Studio Centres: field measurement data.* British Broadcasting Corporation, U.K. October 1986.

6.11. Kryter, K.D. Presbycusis, sociocusis and nosocusis. *J.Acoust.Soc.Amer.* 73, 1983, pp 1987-1917.

6.12. Henderson, D. & Hamernik, R.P. Impulse noise: critical review. *J.Acoust.Soc.Amer.* 80, 1986, pp 569-584.

6.13.—*Acoustics—Assessment for occupational noise exposure for hearing conservation purposes.* ISO 1999. International Standards Organisation.

6.14.—*Acoustics—Hearing Conservation.* AS 1269- 1983, Standards Association of Australia.

6.15. Schroeder, M.R. Binaural dissimilarity and optimum ceilings for concert halls: More lateral sound diffusion. *J.Acoust.Soc.Amer.* 65, 1979, pp 958-963.

6.16. Brebeck, von D., Bucklein, R., Krauth, E. & Spandock, F. Acoustically similar models as auxiliary means in room acoustics. *Acustica* 18, 1967, pp 213-226.

6.17. Burd, A.N. Acoustic modelling—design tool or research project. *Int.Symp on Arch. Acoust.* Edinburgh, 1974.

6.18. Els, H. & Blauert, J. Measuring techniques for acoustic models—upgraded *Internoise* 85, Munich, 1985, pp 1359-1362.

6.19. Heymnan, D., Vermeir, G. & Myncke, H. Comparison between the room responses in real scale and those predicted by a numerical model. *11th Int.Congress on Acoust.*, Paris, 1983, 7, pp 121-124.

6.20. Parkin, P.H., Humphreys, H.R. & Cowell, J.R. *Acoustics, Noise and Buildings*, Faber & Faber, 4th ed. 1979, p.56.

6.21. Houtgast, T. & Steeneken, H.J.M. The modulation transfer function in room acoustics. *Bruel & Kjaer Technical Review*, 3, Copenhagen, 1985, pp 3-12.

6.22. Steeneken, J.J.M. & Houtgast, T. RASTI; A tool for evaluating auditoria. *Bruel & Kjaer Technical Review*, 3, Copenhagen, 1985, pp 13-39.

6.23. Peutz, V.M.A. Variable acoustics of the IRCAM concert hall in Paris, *10th Int. Congress. on Acoust.*, Sydney, 1980, E 1.3.

6.24. Lawrence, Anita. Multi-purpose auditoria—an alternative approach. *10th Int. Congress. on Acoust.* Sydney, 1980, E-6.3.

6.25. Gilford, C. *Acoustics for radio and television studios*, Peter Peregrinus, London, 1972.

6.26. Shankland, R.S. Acoustics of Greek theatres. *Physics Today*, 26, 1973, pp 30-35.

6.27. Cremer, L. The different distributions of the audience, *App.Acoust.* 8, 1975, pp 173-191.

6.28. Parkin, P.H., Humphreys, H.R. & Cowell, J.R. *Acoustics, Noise and Buildings*, Faber & Faber, 1979, p.126.

APPENDIX A

Measurement of the Sound Absorption Coefficients of Materials

A.1 INTRODUCTION

In Section 4.6, the sound absorption characteristics of different building materials and systems were discussed. There are two recognised methods of measuring sound absorption coefficients. The simpler, and therefore cheaper, method uses an *impedance tube*;[A.1.1, A.1.2] the alternative requires the use of a special *reverberation chamber*.[A.1.3, A 1.4]

A.2 IMPEDANCE TUBE METHOD

This method requires a rigid tube, either circular or square in cross section, with a sound source (loudspeaker) at one end and a mounting system for the material sample at the other end. A microphone is mounted on a movable probe tube in order to be able to sample the sound field at different distances from the sample when excited by a single frequency. A standing wave is produced when the plane waves transmitted from the loudspeaker interact with the reduced amplitude waves reflected from the specimen. The relative pressure amplitudes of the maxima and minima of the standing wave pattern are measured, and in addition it is necessary to know, from a calibrated scale, the location of the microphone with respect to the face of the specimen.

The normal incidence sound absorption coefficient, α_n, is determined by measuring the difference between the maximum and minimum sound pressure levels at the surface of the sample, and using the following equation:

$$\alpha_n = 1 - \{[10^{(L_{max} - L_{min})/20} - 1]/[10^{(L_{max} - L_{min})/20} + 1]\}^2 \qquad [A.1.1]$$

where

L_{max} = pressure maximum, dB re 20 μPa
L_{min} = pressure minimum, dB re 20 μPa

Since this coefficient refers only to normal incidence of sound, it is necessary to estimate the random incidence absorption coefficient, α_{stat}, from the measured values of $L_{max} - L_{min}$ and d_1, the distance of the first pressure minimum from the face of the specimen, normalised by the wavelength of the sound, λ. In addition, it is possible to calculate the specific normal acoustic impedance from the measurements. Charts are provided for ease of determination.

Tubes normally used in practice have quite small diameters, of approximately 100 mm, thus this method is only suitable for locally reacting materials—for example, mineral wool, glass fibre and open cell foams; it may also be used to estimate the absorption of perforated panel type materials. The absorption characteristics of these materials are affected by the presence or absence of an air space, and its depth, between the material and the rigid backing of the mounting.

The tube method cannot be used for resonant type absorbers, whose characteristics are affected by panel size.

It can be seen from the above that the impedance tube method of measuring sound absorption coefficients does not subject the material to practical, random incidence sound fields, and thus the results should be used with some caution. However, it is a useful method to use for developmental work.

A.3 REVERBERATION ROOM METHOD

This method requires the reverberation time to be measured in a reverberation room with and without a sample of the material present. In general, the amount of sound energy absorbed by a material depends on the angle of incidence of the sound wave and thus the amount of absorption provided by a material in practice depends on the sound field to which it is exposed. In the reverberation room method, great care is taken to develop a diffuse sound field, i.e. one in which the sound waves will reach the sample from all possible directions, or at *random incidence*.

It has been found that even well-equipped laboratories, measuring the absorption of identical samples, will not achieve exactly similar results. However, the standardised methods attempt to reduce the variation.

It is recommended that the volume of the reverberation room is approximately 200 m^3, and that the reverberation time, empty, is of the order of 5 s in the low frequencies and 2 s in the high frequencies. The material sample should be approximately 10 m^2 area, if it is a planar absorber, and rectangular in shape, with a ratio of width to length between 0.7 and 1. The perimeter of the sample should be enclosed by a frame, to avoid absorption of sound by its edges. This method may also be used to determine the absorption of items such as chairs, people, etc.

Since air absorption is dependent on temperature and relative humidity, it is necessary either to ensure that these parameters do not change during the measurement period, or to make corrections for this effect.

Either broad-band or one-third octave band sound is radiated by one or more loudspeakers in the room, over the range from 100 Hz to 5 kHz. The reverberation times are measured several times for at least three microphone positions for each frequency band.

Assuming that the air attenuation coefficient is constant throughout the measurements, the basic equations used to determine the sound absorption coefficients are as follows:

$$A_e \doteq 0.161 \ V/T_{60,e} \tag{A.1.2}$$

where

> A_e = total equivalent absorption area of empty room, m^2
> V = volume of the room, m^3
> $T_{60,e}$ = the reverberation time of the empty room, s

$$A_{e+s} = 0.16 \ V/T_{60,e+s} \tag{A.1.3}$$

where

> A_{e+s} = total equivalent absorption area of room containing sample, m^2
> V = volume of the room, m^3
> $T_{60,e+s}$ = reverberation time of the room containing the sample, s

$$A_s = A_{e+s} - A_e \tag{A.1.4}$$

$$\alpha_s = A_s/S \tag{A.1.5}$$

where

α_s = absorption coefficient of sample
A_s = equivalent absorption area of sample, m^2
S = sample area, m^2

It can be seen that this method is much closer to the practical situation than is the impedance tube method; however, it is much more expensive and time-consuming.

REFERENCES

A.1.1 —*Acoustics—method for measurement of normal incidence sound absorption coefficient and specific normal acoustic impedance of acoustic materials by the tube method.* AS 1935-1987. Standards Assoc. of Australia.
A.1.2—*Method of test for impedance and absorption of acoustical materials by the tube method.* ASTM C384. Amer.Soc.for Testing and Materials.
A.1.3 —*Measurement of absorption coefficients in a reverberation room.* ISO R 354. International Standards Organisation.
A.1.4—*Acoustics—measurement of sound absorption in a reverberation room.* AS 1045-1988. Standards Assoc. of Australia.

APPENDIX B

Guide to the Sound Attenuating Properties of Typical Building Elements

Airborne Sound Transmission Loss, dB

Material	Centre frequency of 1/3 octave band, Hz																
	100	125	160	200	250	315	400	500	630	800	1 000	1 250	1 600	2 000	2 500	3 150	4 000
110 mm (4½") brickwork, rendered both sides	31	34	35	36	36	37	38	41	45	50	51	53	55	58	59	60	61
Two leaves 110 mm (4½") brickwork, rendered, 50 mm (2") air cavity	39	40	41	42	43	45	46	50	52	55	57	60	61	63	64	66	68
Dense, hollow concrete blocks, 140 mm (5½") thick, rendered 13 mm (½") both sides	—	33	35	39	38	37	41	41	43	45	48	49	53	56	54	55	59
200 mm (8") lightweight aggregate solid blockwork, 10 mm (⅜") dry wall linings both sides	33	35	36	38	40	41	47	50	52	54	55	55	54	54	54	53	52
110 mm (4½") brickwork, 25 mm (1") cavity, 100 mm (4") timber stud frame, 10 mm (⅜") gypsum board lining both sides (brick veneer construction)	31	31	34	38	41	45	49	54	58	61	61	62	65	67	68	67	68

Element																	
100 mm × 50 mm (4" × 2") timber stud frame lined with 13 mm (½") gypsum board both sides	—	14	14	20	22	24	26	31	34	36	39	41	46	46	42	37	37
65 mm (2½") steel studs, 13 mm (½") gypsum board both sides, 50 mm (2") glass fibre batts in cavity	—	23	28	31	33	39	42	48	50	52	54	57	57	56	48	43	45
65 mm (2½") steel studs, two layers 13 mm (½") fire-resisting gypsum board both sides, joints staggered; 50 mm (2") glass fibre in cavity	—	28	37	38	40	44	45	47	49	51	53	54	53	48	44	46	49
Two isolated leaves each 100 mm (4") timber stud frame lined with 13 mm (½") softboard over 13 mm (½") plasterboard on each face; 50 mm (2") air space between frames	37	39	39	42	44	47	49	52	55	59	62	61	65	66	68	69	70
3 mm (⅛") glass in timber or metal frame, openable	17	15	16	15	17	14	18	19	23	22	22	21	18	18	17	20	21
6 mm (¼") glass in metal frame, openable	21	19	18	20	21	23	25	26	26	26	25	24	23	24	26	27	30
10 mm (⅜") laminated glass in metal frame, fixed	—	25	26	28	30	32	32	33	34	36	36	33	36	36	40	44	46
Two leaves 3 mm (⅛") glass in timber or metal frames, 100 mm (4") air space absorbent reveals	20	18	24	26	30	30	33	37	36	43	50	49	50	46	47	47	48
Hollow core plywood door, 45 mm (1¾") thick, no seals	14	12	11	12	11	14	14	14	16	16	17	16	16	16	15	14	16

(continued)

Airborne Sound Transmission Loss, dB—contd.

Material	\multicolumn{17}{c}{Centre frequency of 1/3 octave band, Hz}																
	100	125	160	200	250	315	400	500	630	800	1 000	1 250	1 600	2 000	2 500	3 150	4 000
Solid core plywood door. 42 mm (1¾") thick, gaskets jambs & head, drop seal at bottom	25	23	24	23	26	26	28	26	25	25	25	26	27	28	29	30	34
Tongued & grooved timber flooring. 200 mm × 50 mm (8" × 2") joists; 10 mm (⅜") gypsum board ceiling	15	17	18	22	22	27	29	30	30	31	34	38	41	44	43	41	42
22 mm (⅞") chipboard on 19 mm (¾") plasterboard over 25 mm (1") glass fibre or mineral wool resilient layer over joists; two layers 12 mm (½") plasterboard ceiling	29	34	35	39	42	45	48	50	52	55	56	58	58	59	59	60	61
16 mm (⅝") sheet flooring on 25 mm (1") screed over 150 mm (6") concrete slab. 13 mm (½") rendered ceiling	38	40	35	42	44	45	49	49	50	53	54	57	57	58	59	60	61
16 mm (⅝") sheet flooring on 50 mm (2") screed on 25 mm (1") mineral wool over 150 mm (6") concrete slab. 13 mm (½") rendered ceiling	39	38	38	40	43	43	45	49	54	56	59	60	62	64	64	67	68
0·63 mm (0·02") perforated metal pan. 10% open. 50 mm (2") glass fibre with 1·6 mm (0·06") sealed metal backing; gasketed at joints and perimeter (room to room via ceilings)	—	25	30	32	38	39	44	45	50	53	57	60	60	69	67	68	64
0·7 mm (0·03") steel decking over 30 mm (1⅛") compressed glass fibre over 50 mm (2") gypsum concrete on 25 mm (1") glass fibre formboard ceiling on metal																	

Impact Sound Transmission, dB

Material	Centre frequency of 1/3 octave band, Hz															
	100	125	160	200	250	315	400	500	630	800	1 000	1 250	1 600	2 000	2 500	3 150
Tongued & grooved timber flooring on 200 mm × 50 mm (8" × 2") joists, 10 mm (⅜") gypsum board ceiling	77	78	78	79	79	80	80	80	79	77	76	73	71	68	68	66
22 mm (⅞") chipboard on 19 mm (¾") plasterboard on battens over 25 mm (1") glass fibre or mineral wool resilient layer over joists, two layers 12 mm (½") plasterboard ceiling	69	65	68	67	65	64	60	57	53	48	43	37	33	29	27	24
16 mm (⅝") sheet flooring on 25 mm (1") screed over 150 mm (6") concrete slab, 13 mm (½") rendered ceiling	57	57	59	61	63	64	64	64	64	64	64	64	64	64	64	61
16 mm (⅝") sheet flooring on 50 mm (2") screed over 25 mm (1") mineral wool over 150 mm (6") concrete slab, 13 mm (½") rendered ceiling	63	65	64	63	62	62	62	62	62	61	61	61	60	60	59	58
Tongued & grooved flooring on 50 mm × 25 mm (2" × 1") battens on mineral wool over 150 mm (6") concrete slab, 13 mm (½") rendered ceiling	59	59	59	58	58	57	56	55	53	51	49	46	44	42	43	40

Note: The majority of the data quoted were obtained from measurements in different laboratories. In practical situations, the sound reduction obtained may be much less than predicted from such measurements. It is particularly dependent on the workmanship, and on the possibility of flanking transmission through associated constructional elements. Any gaps or openings will severely limit the overall sound reduction. Whenever possible, laboratory measurements should be made of the actual construction to be used, and these should be compared with *in situ* measurements made after the building is completed.

APPENDIX C

Typical Sound Absorption Coefficients

Material	Frequency, Hz					
	125	250	500	1 000	2 000	4 000
Rendered or plastered masonry or concrete	0·01	0·02	0·02	0·03	0·03	0·04
Plasterboard ceiling on battens, large air space	0·2	0·2	0·1	0·1	0·04	0·02
Timber flooring on joists	0·1	0·1	0·1	0·1	0·1	0·08
Carpet	0·1	0·1	0·2	0·3	0·5	0·6
Window glass	0·2	0·1	0·1	0·1	0·04	0·02
25 mm (1″) mineral wool, solid backing	0·1	0·3	0·7	0·8	0·9	1·0
50 mm (2″) mineral wool, 180 mm (7″) air space, solid backing	0·6	0·8	0·8	0·8	0·8	0·8
3 to 5 mm (0·1–0·2″) panelling over 25 mm (1″) air space, solid backing	0·3	0·5	0·1	0·1	0·0	0·0
3 to 5 mm (0·1–0·2″) panelling perforated 10% over 25 mm (1″) mineral wool, 25 mm (1″) air space, solid backing	0·1	0·3	0·7	0·9	0·8	0·5
3 to 5 mm (0·1–0·2″) panelling perforated 10% over 50 mm (2″) mineral wool, 25 mm (1″) air space, solid backing	0·3	0·8	0·9	0·8	0·7	0·5

Note: These are approximate random incidence sound absorption coefficients, typical of the type of material cited, as measured in a reverberation chamber. Absorption coefficients are strongly dependent on the actual materials used, the angle of the incident sound and on the method of mounting (particularly important is the flow resistance of porous materials and the depth of the air space behind the absorbent, if any). Whenever possible, the measured values of the particular material to be used should be determined.

Index